UPCO'S

Physical Setting

CHEMISTRY

Frederick L. Kirk
Former Science Teacher
Niskayuna High School
Niskayuna, New York

UPCO-United Publishing Co., Inc.
40 Bailey Street
Coxsackie, New York 12051

Editor/Reviewer

Ernest J. Bosco
Retired Chemistry Teacher
Spring Valley High School
Spring Valley, New York

Evan M. Horowitz
Horace Greeley High School
Chappaqua, New York
Chemistry/Biology Teacher

Jay Horowitz
Retired Chemistry Teacher
Ramapo High School
Spring Valley, New York

The author gratefully acknowledges the following teachers and students who have carefully read and made useful suggestions for this book:
Meagan Hughes, Jill Wright, Joseph Spollen, Sherra McGuire, Lara Murphy, TJ Andrews, Eric Battaglioli, Michelle Moon, Peter Chalfin, Andrew Millspaugh, Allison Keith, Elspeth Edelstein, Bryan Lasher, Ava Mauro, Alexander Kent, Andrew Armenia, Andrew Mack, Karin Blecher, Kayla Eisman, Anthony Guzman, Joanne Chaplek, and Sydney Lauren.

Special recognition for their contributions is accorded to:
Mr. Wayne Kneussle and his students
Saranac High School
Saranac, NY

ISBN 978-0-937323-21-2

1 2 3 4 5 6 7 8 9 0

TABLE OF CONTENTS

CHAPTER 1
ATOMIC CONCEPTS

DEFINITION OF CHEMISTRY

Chemistry is the study of the composition, structure, and properties of matter, the changes which matter undergoes, and the energy accompanying those changes.

ATOMS

As early as the 5th century B.C., Greek philosophers believed matter to be composed of separate, indivisible particles that they called atoms. This "particle theory," however, was not generally accepted until the 18th and 19th centuries, when studies of the nature of chemical change by **John Dalton** helped chemists to develop the **atomic theory** of the structure of matter.

SUBATOMIC PARTICLES

Most of the **volume** of an atom consists of empty space. Most of the **mass** of an atom is concentrated in a dense, centrally located nucleus which contains protons and neutrons. The nucleus carries the positive charge of the atom. The region surrounding the nucleus is occupied by tiny, fast-moving, negatively charged particles called electrons.

Protons are those particles considered to have a mass of one atomic mass unit and one unit of positive (+) charge.

Neutrons are those particles that have a mass about the same as that of protons, but with a charge of zero.

Electrons. The mass of an electron is very small compared to that of a proton. This mass is so small that electrons are generally considered to have no mass at all. An electron has one unit of negative (–) charge that is equal in magnitude (and opposite in electrical nature) to the charge on a proton.

ATOMIC STRUCTURE

The atoms that make up different kinds of matter are different, but their subatomic particles are all alike. Differences in atoms result from differences in the *number* of protons and neutrons in their nuclei and in the *number* and *arrangement* of their electrons.

1

Ernest Rutherford's experiments in the early 1900's demonstrated that most of the space occupied by an atom is empty most of the time.

Atomic Number. Depending on data collected using X-rays, **Henry Moseley** showed that each element possessed a characteristic number of positive charge units (protons). This number is now considered to be the **atomic number** of the element, and it is equal to the number of protons in the nucleus, as well as to the charge on the nucleus.

Atoms of elements are ordinarily electrically neutral, because the number of negative charges (electrons) is *equal* to the number of positive charges (protons).

Mass Number. The total number of protons + neutrons in an atom is called the **mass number**. Since the masses of protons and neutrons are each assigned a value of one atomic mass unit, the mass number of an atom is a whole number.

Isotopes. Every atom of each element has the same atomic number, because they all have the same number of protons. Some atoms of the same element do, however, have different numbers of neutrons. Since neutrons contribute about the same mass as protons, the masses of atoms of the same element can be slightly different. Atoms of elements having the same atomic number but different masses are called **isotopes** of that element. When an isotope is identified by symbol, atomic number, and mass number, it is often referred to as a **nuclide**.

The normal designation for isotopes is as follows (using sodium -23 as an example):

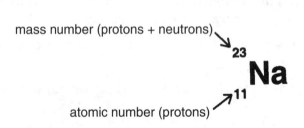

Figure 1-1. Normal designation for isotopes.

Atomic Mass. The **atomic mass** of an element is the term used to express the weighted average mass of the naturally occurring isotopes of that element. This average is weighted according to the ratios in which the isotopes occur. Most elements occur in nature as mixtures of their isotopes.

For example: lithium has two isotopes ^{6}Li and ^{7}Li. 7.4% is ^{6}Li and 92.6% is ^{7}Li.

^{6}Li × 7.4% ^{7}Li × 92.6%
(6.0) × (.074)   (7.0) × (.926)
    0.44   +   6.48   =   6.92 or 6.9 This is the atomic mass.

    The mass number of the most abundant isotope can usually be found by rounding off the atomic mass of the element to the nearest whole number.

Electrons. Electrons are located in *energy levels* that exist at various distances outside the nucleus. In neutral atoms, the total number of electrons is equal to the number of protons in the nucleus (the atomic number of the element).

QUESTIONS

    Answer the following questions using the Periodic Table of the Elements or Table S for atomic numbers and symbols.

1. The number of protons in the nucleus of $^{32}_{15}\text{P}$ is

  (1) 15   (2) 17   (3) 32   (4) 47

2. What is the total number of electrons in an atom with an atomic number of 13 and a mass number of 27?

  (1) 13   (2) 14   (3) 27   (4) 40

3. Which symbol represents an isotope of carbon?

  (1) ^6_4C (2) $^{12}_5\text{C}$ (3) $^{13}_6\text{C}$ (4) $^{14}_7\text{C}$

4. The mass number of an atom is equal to the total number of its

(1) electrons, only   (2) protons, only
(3) electrons and protons   (4) protons and neutrons

5. What is the total number of protons in an atom of $^{36}_{17}\text{Cl}$?

  (1) 17   (2) 18   (3) 35   (4) 36

6. Which nuclide contains the greatest number of neutrons?

  (1) $^{37}_{17}\text{Cl}$ (2) $^{39}_{19}\text{K}$ (3) $^{40}_{18}\text{Ar}$ (4) $^{41}_{20}\text{Ca}$

7. The nucleus of an atom of $_{53}^{127}I$ contains

    (1) 53 neutrons and 127 protons   (2) 53 protons and 127 neutrons
    (3) 53 protons and 74 neutrons   (4) 53 protons and 74 electrons

8. What is the mass number of an atom which contains 21 electrons, 21 protons, and 24 neutrons?

    (1) 21   (2) 42   (3) 45   (4) 66

9. How many electrons are in an atom of beryllium?

    (1) 5   (2) 2   (3) 9   (4) 4

10. If X is the symbol of an element, which pair correctly represents isotopes of X?

    (1) $_{64}^{158}X$ and $_{64}^{158}X$ (2) $_{158}^{64}X$ and $_{64}^{158}X$ (3) $_{64}^{158}X$ and $_{64}^{159}X$ (4) $_{64}^{158}X$ and $_{65}^{158}X$

11. Which two particles have approximately the same mass?

    (1) neutron and electron   (2) neutron and deuteron
    (3) proton and neutron   (4) proton and electron

12. What is the symbol for an atom containing 20 protons and 22 neutrons?

    (1) $_{20}^{42}Ca$ (2) $_{20}^{40}Ca$ (3) $_{22}^{42}Ti$ (4) $_{22}^{40}Ti$

MODELS OF ATOMIC STRUCTURE

When Dalton presented his proposal of atomic structure, he pictured atoms as solid spheres that were indivisible and complete in themselves. **J. J. Thomson's** discovery of the electron required a new model that included both positive and negative parts. Thomson proposed an atom that consisted of a solid, positively charged sphere. The negatively charged electrons were embedded in the surface of the atom.

Rutherford's experiments showed that the electrons were located *outside* the positively charged portion of the atom. His model showed a dense, positively charged nucleus with electrons somewhere outside the nucleus. Since Rutherford's time, the model of the atom has undergone many changes and refinements, and there will probably be many more to come.

In 1913, the Danish physicist, **Niels Bohr**, proposed a model of the atom in which electrons revolved around the nucleus in concentric, circular orbits, like the planets around the Sun.

Principal Energy Levels. Each of the orbits in the Bohr model of the atom has a fixed radius. The greater the radius of an orbit (the farther from the nucleus), the greater the energy of the electrons in the orbit. The orbits, or "shells" of the Bohr model are known as **principal energy levels**, and are denoted either by the letters K, L, M, N, O, P and Q or by the numbers 1 through 7.

In this model, when all the electrons are located in the lowest available energy levels, the atom is said to be in the **ground state**. However, electrons may absorb a specific amount of energy from outside the atom to move to a higher energy level, leaving a lower level unfilled. In this condition, the atom is said to be in the **excited state** and is, therefore, unstable. The excited electron(s) will soon "fall" to the lower level. As this occurs, the excess energy is released in some form, such as light or heat.

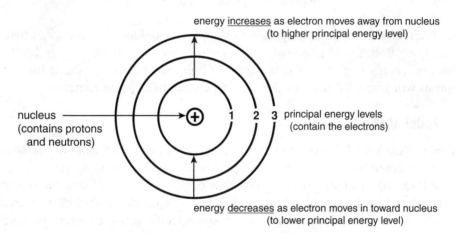

Figure 1-2. Model of an atom.

Quanta. One important feature of the Bohr atom is the idea that electrons can only absorb or release energy in discrete, specific, amounts. These amounts, or "bundles" of energy, called **quanta** (singular, quantum), correspond to the differences in energy levels of the shells.

Spectral Lines. If high voltage is applied to hydrogen gas confined in a glass tube, called a gas discharge tube, light is emitted. If this light is passed through a prism, a series of bright lines of distinct colors is produced. These lines can be projected onto a screen. Bohr reasoned that these different colored bands of light were actually quanta of correspondingly different energy. These quanta were emitted as electrons of the hydrogen atoms returned from their higher levels in the excited state to their lower levels in the ground state.

The device designed to observe the separation of light into its component colors (wavelengths) is called a spectroscope. A spectroscope consists mainly of a prism and a telescope. The series of bright lines produced when excited electrons return to their original energy levels is called a **bright-line spectrum.** Each element has its own unique set of spectral lines which can therefore be used to identify the element's presence.

Below are the visible bright line spectras of five elements and the continuous spectrum of a white light light bulb.

Continuous Spectrum	red	orange	yellow	green	blue	violet
H						
He						
Na						
Cd						
Li						

Figure 1-3. Spectral diagram.

The blue spectral line for hydrogen, for example, is produced by an excited hydrogen electron dropping from the 4th principal energy level to the 2nd principal energy level. This loss in electron energy is given off in the form of light which, in this case, is blue. Other electron drops will give off different colors corresponding to different energies.

Orbital Model of the Atom

Although Bohr's model of the atom provided an explanation for the bright-line spectrum of hydrogen, it could not account for the spectra of atoms containing many electrons. Scientists, influenced by studies of the wave behavior of electrons, replaced the Bohr model with one that describes the motion of electrons in terms of the *probability* of their positions within the atom. This **orbital model** does not assign specific paths, or orbits, in which the electrons move. Rather, it considers the electrons to move freely around the nucleus. The regions of most probable electron location may differ in size, shape, and orientation in space. These regions are known as **orbitals**.

A cross section of two orbitals are shown below. These electron clouds represent the most probable locations of finding a tiny electron in an atom. The nucleus of the atom (+) is at the center of each orbital.

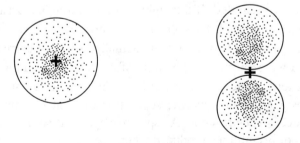

Figure 1-4. Electron clouds.

The probability is highest for finding the electron where the dots are most dense. (The shapes of orbitals are not New York State Regents testable.)

Electron Configurations. There are several different methods for showing, in symbolic form, the distribution of electrons of the atoms of each of the elements in the ground state. These symbolic devices are called **electron configurations**. The following set of rules is the basis for showing these electron distributions:

1.   The number of electrons must equal the atomic number.

2.   Each principal energy level can hold the following maximum number of electrons:

1st	-	2
2nd	-	8
3rd	-	18
4th	-	32
5th	-	50

3.   Electrons generally occupy energy levels in sequence, beginning with those of lowest energy.

4.   No more than eight electrons occupy the outermost principal energy level (except that the 1st can only hold two).

5.   The 4th principal energy level gets two electrons before the 3rd gets electrons 9-18. (Similar irregularities occur at higher principal energy levels - see your *Reference Tables for Physical Setting/Chemistry - Periodic Table of the Elements*).

element	electron configuration	explanation
Al	2-8-3	2 electrons in the 1st, 8 electrons in the 2nd, 3 electrons in the 3rd energy level
K	2-8-8-1	2 electrons in the 1st, 8 electrons in the 2nd, 8 electrons in the 3rd, 1 electron in the 4th energy level
Sc	2-8-9-2	see rule 5 above

Figure 1-5. Some elements and their electron configuration notation.

Valence Electrons. The electrons that are most responsible for the properties and chemical reactivity of the elements are those in the outermost principal energy level. These electrons are known as **valence electrons.** There cannot be more than eight valence electrons.

Sometimes it is useful to show the distribution of valence electrons by means of diagrams called **electron-dot symbols.** In an electron-dot symbol, the kernel (nucleus plus non-valence electrons) of the atom is represented by the chemical symbol of the element. The valence electrons are represented by dots arranged around the kernel. These dots are usually placed in the "clock" positions: 12 o'clock, 3 o'clock, 6 o'clock, and 9 o'clock.

Electron-dot diagrams for the first 18 elements are shown in the following illustration:

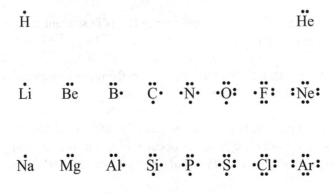

Figure 1-6. Electron dot diagrams.

QUESTIONS

Use the Periodic Table of the Elements and Table S for reference.

1. Which term refers to the region of an atom where an electron is most likely to be found?

(1) orbital   (2) nucleus   (3) quantum   (4) spectrum

2. An atom has the electron configuration 2-8-7. The electron dot symbol for this element is

(1) X⦂   (2) Ẋ⦂   (3) ·Ẋ⦂   (4) ⦂Ẍ⦂

3. What is the total number of electrons in the second principal energy level of a calcium atom in the ground state?

   (1) 6   (2) 2   (3) 8   (4) 18

4. As an electron in an atom falls to a lower energy level, the energy of the electron

   (1) decreases   (2) increases   (3) remains the same

5. Which principal energy level change by the electron of a hydrogen atom will cause the greatest amount of energy to be absorbed?

   (1) 2 to 4   (2) 2 to 5   (3) 4 to 2   (4) 5 to 2

6. Spectral lines produced from the energy emitted from excited atoms are thought to be due to the movements of electrons

   (1) from lower to higher energy levels   (2) from higher to lower energy levels
   (3) within their nuclei   (4) out of the nucleus

7. The diagram below represents the electron dot notation of an atom's valence shell in the ground state.

$$\ddot{X}\cdot$$

   The diagram could represent the valence shell of

   (1) Li   (2) Si   (3) Al   (4) Cl

8. The principal energy level of the outermost electron of an atom in the ground state is 3. What is the total number of occupied principal energy levels contained in this atom?

   (1) 1   (2) 2   (3) 3   (4) 4

9. How many occupied energy levels are in an atom of copper in the ground state?

   (1) 5   (2) 6   (3) 3   (4) 4

10. What is the total number of valence electrons in an atom with the electron configuration 2-8-5?

   (1) 5   (2) 11   (3) 3   (4) 15

11. Which electron transition is accompanied by the emission of energy?

(1) 1st to 2nd (2) 2nd to 1st (3) 3rd to 5th (4) 3rd to 3rd

12. The maximum number of electrons that the 4th energy level may contain is

(1) 2   (2) 8   (3) 18   (4) 32

13. Which is the electron configuration of an atom in the excited state?

(1) 2-5   (2) 2-8-8-1   (3) 2-9-8-2   (4) 2-8-7-1

14. Which is the electron dot symbol of an atom of boron in the ground state?

(1) $\cdot\overset{\bullet}{\underset{\bullet}{B}}\colon$ (2) B$\cdot$ (3) $\cdot\overset{\bullet}{B}\colon$ (4) $\overset{\bullet}{B}\colon$

15. Which atom has a completely filled 3rd principal energy level?

(1) Ar   (2) Zn   (3) Ca   (4) K

16. Which is the atomic number of an atom with six valence electrons?

(1) 6   (2) 8   (3) 10   (4) 12

PRACTICE FOR CONSTRUCTED RESPONSE

1. Below are bright line spectra of five elements, a continuous white light spectrum with its colors, and the spectrum of an unknown gas X.

Continuous Spectrum	red	orange	yellow	green	blue	violet
H						
He						
Na						
Cd						
Li						
Unknown X						

    a. Which two elements are in unknown X? (2 points)

_____ and _____

    b. Is there a third element present in unknown X? Explain your answer. (2 points)

_____

    c. Stars are mostly hydrogen with a small abundance of other elements. Is unknown X a likely spectrum for a Star? Explain your answer. (2 points)

_____

2. For a neutral atom of $^{7}_{3}\text{Li}$

    a. How many protons, neutrons, and electrons are there? (3 points)

protons _____ # neutrons _____ # electrons _____

    b. How many energy levels have electrons in the ground state? (1 point)

_____

c. Sketch a model of this atom with the protons, neutrons, and electrons labeled, and put them in their proper locations. (3 points)

3. An imaginary element X has just two isotopes, ^{42}X and ^{44}X. 20% of all atoms of X are ^{42}X.

a. Calculate the atomic mass of X. Show your work. (2 points)

_____

b. Why will the atomic mass never be 42.0 or 44.0 if determined by laboratory experiment? (1 point)

_____

4. The electron configuration 2-8-18-7-3 is for a neutral atom.

a. What is the name of the element? (1 point)

_____

b. Is the atom in the ground state or the excited state? (1 point)

_____

c. Will the atom spontaneously absorb or release energy? Explain your answer. (1 point)

_____

_____

d. Draw the electron dot formula for this atom if it were in the ground state. (1 point)

5. Match the scientist with the model of the atom he founded. (4 points)

(1) Ernest Rutherford　　　　　(2) Niels Bohr

(3) J. J. Thomson　　　　　　　(4) John Dalton

a. Atoms are solid, indivisible particles.　　　　　_____

b. Electrons are arranged in energy levels in orbit around the nucleus.　_____

c. An atom is a solid positive particle with electrons imbedded within it.　_____

d. An atom has a positive nucleus at the center of the atom with electrons outside the nucleus.　_____

6. In the box below the name of the scientist, sketch the model of the atom which each developed. (4 points)

John Dalton　　　J. J. Thomson　　　Ernest Rutherford　　　Niels Bohr

7. Why is the "17" in $^{35}_{17}Cl$ not needed to describe chlorine?

_____

_____

CHAPTER 2
PERIODIC TABLE

DEVELOPMENT OF THE PERIODIC TABLE

Classification is a very useful device for dealing with information. The *Periodic Table of the Elements* evolved into its present form as a result of several schemes for classifying the elements. Early in the development of the table, **Dmitri Mendeleev** and others considered regularities in properties of the elements to be related to their atomic masses. **Henry Moseley** has since established that the properties of elements depend on the structure of their atoms and vary with the atomic number in a more systematic way.

Groups. The vertical columns of the periodic table are called **groups** or **families** of elements. They are designated by the numerals 1-18. Group 2, for example, consists of Be, Mg, Sr, Ba and Ra. The elements in each group exhibit similar or related properties.

Periods. The horizontal rows of the periodic table are called **periods**, or **rows**. They are designated by the numerals 1-7. Period 3, for example, consists of Na, Mg, Al, Si, P, S, Cl, and Ar. The number of the period is the same as the number of the outermost principal energy level containing electrons. The elements in period 3 have their outermost (valence) electrons in the *third* principal energy level. The properties of the elements in a given period tend to vary in a systematic way.

PROPERTIES OF ELEMENTS IN THE PERIODIC TABLE

Attraction For Electrons

The attraction of an atom for its electrons can be measured in several ways. Two ways are *ionization energy* and *electronegativity*, both of which appear on Table S of the *Reference Tables for Physical Setting/Chemistry.*

Ionization energy is the least energy needed to remove an electron from an atom. The more an atom attracts its electrons, the more energy it takes to remove one of them.

Electronegativity is the relative scale which measures the attraction an atom has for electrons shared in a chemical bond (the bond between atoms). The more an atom attracts its bonding electrons, the greater its electronegativity.

Atomic Radius

The size of an atom and its attraction for electrons are related. The size (radius) of an atom is determined by how strongly an atom's nucleus (positive charge) is attracted to its outermost electrons (negative charge). This attraction depends on three factors.

1. The **nuclear charge** (number of protons) attracts the outermost electrons. An increased nuclear charge will attract electrons more, pulling the electrons in closer.
2. The **principal energy level** affects the distance from the nucleus. The higher the principal energy level of the outermost electrons, the farther they are from the nucleus.
3. The **electron cloud effect** shields the outermost electrons from the nucleus. The effect is caused by the inner electrons repelling the outermost electrons because they have the same charge. The greater the number of inner electrons, the less strongly the outermost electrons are attracted to the nucleus and the larger the size of the atom.

The radius of an atom is the closest distance to which it will approach another atom of any size under the circumstances specified. The atomic radius is a periodic property of the elements.

Within a single period of the periodic table, the atomic radius generally decreases as the atomic number increases.

The members of any group in the periodic table generally show an increase in atomic radius with an increase in atomic number.

Metals

Most of the elements (more than two-thirds) are metals. Their atoms lose electrons fairly easily because of their low ionization energies. The resulting ions are positively charged. Metals, therefore, are sometimes referred to as *electropositive elements*. Their properties include:

1. low ionization energy and electronegativity
2. formation of positive ions when combining with other atoms by losing electrons.
3. being solids at room temperature (except mercury, Hg, which is a liquid).
4. shining with a metallic luster when polished.
5. being good conductors of heat and electricity.
6. being able to be rolled or hammered into sheets easily. This property is called *malleability.*
7. can be drawn (pulled) into wire easily. This property is called *ductility.*

Elements with the most pronounced metallic properties appear in the lower left region of the periodic table.

Nonmetals

The atoms of nonmetals gain electrons easily, producing ions with negative charges. They have high attractions for electrons. Other properties include:

1. high ionization energies and electronegativities.
2. forming negative ions when combining with metal atoms by gaining electrons.
3. production of covalent bonds by sharing electrons with other nonmetals.
4. existing as gases, molecular solids, or network solids at room temperature (except for bromine, which is a volatile liquid).
5. being brittle solids, possessing little or no malleability or ductility.
6. being dull solids, even when polished.
7. being poor conductors of heat and electricity.
8. being solids, liquids or gases at room temperature (Bromine, Br_2, is the only one which is a liquid).
9. many being diatomic molecules (e.g. F_2, Cl_2, Br_2, I_2, At_2, N_2, O_2, H_2).

The most pronounced nonmetallic elements appear in the upper right region of the periodic table, except for the elements in Group 18 (the noble gases).

Metalloids (semi-metals)

Elements that appear at the border between the metals and the nonmetals have some of the properties of both. As metals and non-metals they are called *metalloids* or *semi-metals*. They can be found along the heavy black step line of the Periodic Table of the Elements. Included among the metalloids are the elements boron, silicon, germanium, arsenic, antimony, and tellurium.

Noble Gases

There is a small group of gases on the right edge of the Periodic Table of the Elements. They are called the *noble gases,* and they have very different chemical properties from the other elements. These properties include:

1. being chemically unreactive.
2. having completely filled valence energy levels.
3. being, without exception, monatomic.

QUESTIONS

Use the Periodic Table of the Elements as well as Table S in answering the following qustions.

1. Compared to a neon atom, a helium atom has a

(1) smaller radius   (2) smaller first ionization energy
(3) larger atomic number   (4) greater number of electrons

2. Compared to atoms of metals, atoms of nonmetals generally have

(1) higher electronegativities and lower ionization energies
(2) higher electronegativities and higher ionization energies
(3) lower electronegativities and lower ionization energies
(4) lower electronegativities and higher ionization energies

3. Which element is an active nonmetal?

(1) radon   (2) oxygen   (3) zinc   (4) chromium

4. Which properties are characteristic of nonmetals?

(1) low thermal conductivity and low electrical conductivity
(2) low thermal conductivity and high electrical conductivity
(3) high thermal conductivity and low electrical conductivity
(4) high thermal conductivity and high electrical conductivity

5. The elements in the present Periodic Table of the Elements are arranged according to their

(1) atomic numbers   (2) atomic masses
(3) mass numbers   (4) oxidation states

6. A property of most nonmetals in the solid state is that they are

(1) brittle   (2) malleable
(3) good conductors of electricity   (4) good conductors of heat

7. On the Periodic Table of the Elements, an element classified as a semimetal (metalloid) can be found in

(1) Period 6, Group 15,   (2) Period 2, Group 14
(3) Period 3, Group 16   (4) Period 4, Group 15

8. Which element listed below has the *least* metallic character?

(1) Na   (2) Mg   (3) Al   (4) Si

9. An atom of chlorine and an atom of bromine have the same

(1) electronegativity   (2) ionization energy
(3) radius   (4) number of valence electrons

THE CHEMISTRY OF A GROUP (FAMILY)

The main reason for the similarity or relatedness of the properties of the elements in a given Group, or Family, with respect to chemical activity, is that *all members have the same numbers of valence electrons.* Typifying this similarity is their behavior as they combine with other elements. The types of compounds produced are very much the same. Group 1 metals form chlorides with the general formula MCl, with M representing any member of the group. The metals in Group 2 form chlorides with the general formula MCl_2. There is a progressive change in the properties of the metals in a group as their atomic numbers increase. Examples include: the covalent atomic radius of an atom increases as the atomic number increases; the ionization energy and electronegativity of an atom generally decreases, as the distance between the nucleus and valence electrons increases. This latter change is also the result of the larger electron cloud effect (the inner electrons repelling the outer electrons) in which more energy levels are involved. The inner electrons seem to form a barrier, or "shield," against the attraction between the nucleus and valence electrons.

Groups 1 and 2

Group 1 contains the elements known as the *alkali metals* which have one valence electron. Group 2 contains the elements known as the *alkaline earth metals* which have 2 valence electrons. The valence electrons of both groups are easily lost, making these elements highly reactive. As a result, these elements can be found in nature only in compounds. In most cases, the compounds are processed by electrolytic methods in order to liberate the elemental metals. (See p. 149)

Elements in both groups 1 and 2 exhibit low ionization energies and low electronegativities, and readily form very stable ionic compounds. The reactivity of both the alkali metals and the alkaline earth metals increases as we proceed down a group but this property tends to decrease as the atomic number increases within a given

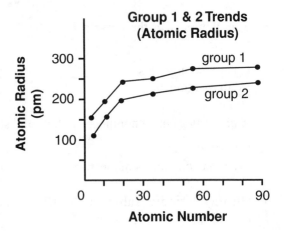

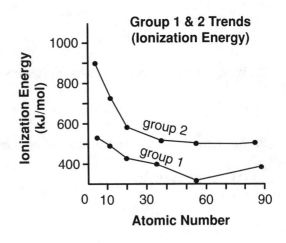

period. For example, in Period 3, sodium in Group 1 is more reactive than magnesium in Group 2. In this case, the increased reactivity is attributed to the larger atomic sizes in Group 1, when compared to those in Group 2, and also the greater nuclear charge in the elements of Group 2.

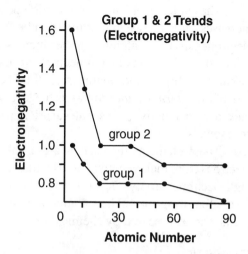

Group 16

Elements in Group 16 have six valence electrons. They show a progression of properties from nonmetallic to metallic as atomic number increases. The first two members of the group, oxygen and sulfur, have typical nonmetallic properties. Selenium and tellurium are metalloids, and the last member, polonium, is a metal. With the exception of oxygen, which is a diatomic gas at room temperature, the elements in Group 16 are solids. As an element, oxygen forms compounds with most other elements fairly easily. In combining with other elements, oxygen exhibits a –2 oxidation state in most cases, and it has a high electronegativity. However, oxygen will combine with fluorine, which is more electronegative. Oxygen behaves like a metal in this case and demonstrates a +2 oxidation state. The abundance of oxygen in the free state, in spite of its high reactivity, is the result of its being the product of photosynthesis in plants.

The lower reactivity of sulfur accounts for its abundance in its elemental state. This lower reactivity is undoubtedly related to its higher atomic number. Sulfur exhibits both negative and positive oxidation states when it combines with other elements.

Although selenium and tellurium react with hydrogen, with the anticipated negative oxidation state, they usually produce compounds by virtue of positive oxidation states.

The last element in the group, polonium, is highly *radioactive*. It is produced as a decay product of radioactive uranium.

Group 17

The elements of Group 17 have seven valence electrons. They are known as the *halogens*. They include fluorine, chlorine, bromine, iodine, and astatine. As with all the nonmetals, the metallic character of the halogens increases with the atomic number. Each of the halogens has a high electronegativity, with fluorine being the most electronegative of *all* the elements. As a result, fluorine exhibits a negative oxidation state in *all* its compounds. The other members of the group demonstrate positive oxidation states when they form compounds with more electronegative elements.

The tendency of halogens to exhibit positive oxidation states when forming compounds increases with increasing atomic number. This is in keeping with the trend in decreased electronegativity as atomic number increases. In the free state, halogens occur as **diatomic molecules** (two atoms in a molecule). At room temperature, fluorine and chlorine are gases, bromine is a liquid, and iodine is a solid. The differences in phase are related to the increasing size of the *intermolecular forces* as the size of their molecules increases. (See p. 56) Because they are so reactive, it is not surprising to find that halogens occur in nature only as compounds.

These elements themselves are usually extracted from their compounds by removing electrons from their negative ions (halides). Fluorine is the exception because of its extremely high electronegativity. The only way to prepare free fluorine is by melting its compounds, and then submitting the melted compounds to electrolytic processes. Chlorine, bromine, and iodine can be prepared by chemical means.

Group 18

The elements of Group 18 have eight valence electrons. They are known as the **noble gases**. They were in the past referred to as the inert gases because, it was once thought to be impossible to produce compounds with any of them. However, in recent years, compounds of krypton, xenon, and radon have been produced with the elements fluorine and oxygen. The inability of Group 18 elements to react with other elements is attributed to the stability produced by their electron configurations, in which the outer shells are completely filled.

Groups 3-11

The *transition elements*, which are found in the central region of the periodic table, generally exhibit positive oxidation states. They can often form ions of more than one stable charge. Iron can be Fe^{2+} or Fe^{3+}, and copper can be Cu^+ and Cu^{2+}, for example. This is because many transition metals can sometimes lose electrons from their highest non-valence energy level as well as from their valence energy level. Compounds containing transition metals often have bright colors.

QUESTIONS

Use the Periodic Table of the Elements as well as Table S in answering the following questions.

1. The reactivity of the metals in Groups 1 and 2 generally increases with
   (1) increased ionization energy   (2) increased atomic radius
   (3) decreased nuclear charge   (4) decreased mass

2. The alkaline earth metals are found in Group
   (1) 1   (2) 2   (3) 11   (4) 12

3. Which is the most active nonmetallic element in Group 16?

    (1) oxygen   (2) sulfur   (3) selenium   (4) tellurium

4. Which of the Group 15 elements can lose an electron most readily?

    (1) N   (2) P   (3) Sb   (4) Bi

5. Which element in Group 15 has the greatest metallic character?

    (1) N   (2) P   (3) Sb   (4) Bi

6. In which group are all the elements found naturally only in compounds?

    (1) 18   (2) 2   (3) 11   (4) 14

7. Which three elements have the most similar chemical properties?

    (1) Ar, Kr, Br   (2) K, Rb, Cs   (3) B, C, N   (4) O, N, Si

8. Which sequence of atomic numbers represents elements which have similar chemical properties?

    (1) 19, 23, 30, 36   (2) 9, 16, 33, 50   (3) 3, 12, 21, 40   (4) 4, 20, 38, 88

9. Which groups contain metals that are so active chemically that they occur naturally only in compounds?

    (1) 1 and 2   (2) 2 and 12
    (3) 1 and 11   (4) 11 and 12

10. In the ground state, how many electrons are in the outermost energy level of each element in Group 17?

    (1) 5   (2) 2   (3) 7   (4) 8

11. Which halogen has the least attraction for electrons?

    (1) F   (2) Cl   (3) Br   (4) I

12. As the elements in Group 2 are considered from beryllium to radium, the degree of metallic activity

    (1) increases and atomic radius increases
    (2) increases and atomic radius decreases
    (3) decreases and atomic radius increases
    (4) decreases and atomic radius decreases

13. Which group of elements in the Periodic Table of the Elements contains a semimetal (metalloid)?

(1) 1   (2) 7   (3) 13   (4) 18

14. Which element under normal conditions exists as monatomic (one atom) molecules?

(1) N   (2) O   (3) Cl   (4) Ne

15. Oxygen has a positive oxidation number when it is in a compound with

(1) I   (2) Br   (3) Cl   (4) F

16. Which group contains elements with a total of four electrons in the outermost principal energy level?

(1) 1   (2) 18   (3) 16   (4) 14

17. Which of the following Group 17 elements has the highest melting point?

(1) fluorine   (2) chlorine   (3) bromine   (4) iodine

18. Which element is a liquid at room temperature?

(1) K   (2) I_2 (3) Br_2 (4) Mg

19. As the elements in Group 18 are considered in order of increasing atomic number, the ionization energy of each successive element

(1) decreases   (2) increases   (3) remains the same

20. Elements whose two outermost sublevels may be involved in a chemical reaction are called

(1) noble gases   (2) halogens
(3) alkali metals   (4) transition metals

21. Which atom has multiple oxidation states and forms an ion that is colored when in solution?

(1) Cl   (2) F   (3) Cu   (4) Zn

22. Which element may react chemically by losing electrons from two principal energy levels?

    (1) Al   (2) Cl   (3) Fe   (4) Be

23. The water solution of a compound is bright yellow. The compound could be

    (1) KNO_2 (2) K_2CrO_4 (3) KOH   (4) K_3PO_4

24. Which compound is colorless in a water solution?

    (1) $Al_2(SO_4)_3$ (2) $Cr_2(SO_4)_3$ (3) $Fe_2(SO_4)_3$ (4) $Co_2(SO_4)_3$

25. As the atoms of the elements in Group 1 are considered in order from top to bottom, compared to the ionization energy of the atom above it, the ionization energy of each successive atom

    (1) decreases   (2) increases   (3) remains the same

26. The valence electrons represented by the electron dot $\cdot \overset{\cdot\cdot}{X} \cdot$ symbol could be those of atoms in Group

    (1) 13   (2) 15   (3) 3   4) 16

THE CHEMISTRY OF A PERIOD

In each period of the periodic table, the elements have the same number of principal energy levels. Their valence electrons are in the same principal energy level. In a given period, the properties of the elements change systematically as their atomic numbers increase.

As the atomic number increases:

1. the atomic radius decreases.
2. there is an increase in the ionization energy and electronegativity.
3. there is a general change in character from that of very active metallic elements to that of the less active nonmetallic elements, to very active nonmetals, and, finally, to that of noble gases.
4. there is a gradual change from positive to negative oxidation states.
5. the metallic characteristics of the elements decrease. It should be pointed out that this trend does not apply to the transition elements.

Period 2 & 3 Trends
(Atomic Radius)

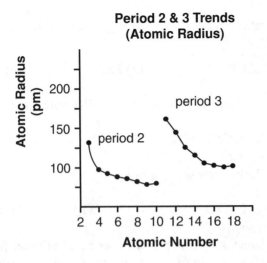

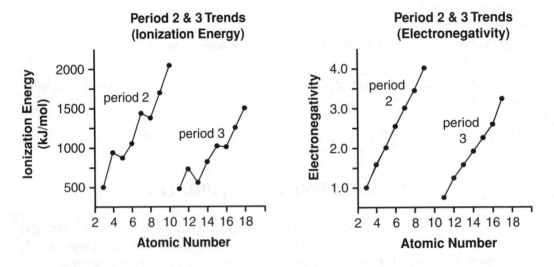

QUESTIONS

Use the Periodic Table of the Elements and Table S to answer the following questions.

1. Of all the elements, the one with the highest electronegativity is found in Period

(1) 1   (2) 2   (3) 3   (4) 4

2. Which element exists as a diatomic molecule under normal conditions (STP)?

   (1) bromine   (2) argon   (3) sulfur   (4) rubidium

3. As the elements of Period 2 are considered in succession from left to right, there is a general decrease in

   (1) ionization energy   (2) electronegativity
   (3) metallic character   (4) nonmetallic character

4. Which element in Period 3 has the most metallic character?

   (1) Al   (2) Si   (3) Na   (4) Mg

5. Which element in Period 4 of the Periodic Table of the Elements exhibits the most nonmetallic properties?

   (1) Ca   (2) Cr   (3) Ga   (4) Br

6. The highest ionization energies in any period are found in Group

   (1) 1   (2) 2   (3) 17   (4) 18

7. Which element in Period 2 has the greatest tendency to gain electrons?

   (1) Li   (2) C   (3) F   (4) Ne

8. Elements in Period 3 are alike in that they all have the same number of

   (1) protons   (2) neutrons
   (3) electrons in the valence shell   (4) occupied principal energy levels

9. Which element in Period 3 is the most active nonmetal?

   (1) sodium   (2) magnesium   (3) chlorine   (4) argon

10. Which of the following electron configurations represents the element with the *smallest* radius?

    (1) 2-4   (2) 2-5   (3) 2-6   (4) 2-7

PRACTICE FOR CONSTRUCTED RESPONSE

1. An abbreviated Periodic Table of the Elements is shown below. All elements are chemically reactive , and A-H will be the symbols this question will use for some elements. (6 points - 1 point each)

Groups

A		E	G
	C		
	D		
B			H
		F	

Periods

   a. Which element has the highest electronegativity?   _____

   b. Which element is most metallic?   _____

   c. Which element has the greatest atomic mass?   _____

   d. Which element is most chemically similar to A: _____

   e. Which element has the smallest atomic radius?   _____

   f. Explain how you know that none of the elements shown are a noble gas?

_____

_____

2. The properties of five elements, V, W, X, Y and Z, are listed below:

        V - a gas, does not react chemically with any element.

        W - a solid, high density, good conductor of electricity.

        X - brittle solid, high electronegativity, poor conductor of heat and electricity.

        Y - brittle solid, metallic luster, conducts electricity somewhat.

        Z - a gas, reacts with sodium to form Na_2Z.

    a. Which is a metal? _____

    b. Which is a non-metal? _____

    c. Which is a metalloid? _____

    d. Which is a noble gas? _____

    e. Which is most likely to react chemically with element Z to form
       an ionic bond?
       (1 point each part) _____

3. Using Table S of the *Reference Tables for Physical Setting/Chemistry,*

    a. On the left axis below, plot the ionization energies of the halogens (Group 17) as a
       function of their atomic number . (2 points)

    b. On the right axis below, plot the ionization energies of the Period 6 elements #83–86
       as a function of their atomic numbers. (2 points)
       (Be sure to label your axes on both graphs.)

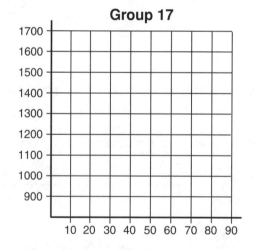
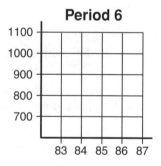

    c. Based on your graphs, predict a reasonable value for the ionization energy of astatine
       (atomic number 85). (1 point)

_____

    d. Explain how one could arrive at an answer for part c. above. (1 point)

_____

_____

_____

4. A compound has a formula of XCl_2.

    a. What Group on the Periodic Table of the Elements is X in? (1 point)

    _____

    b. How many valence electrons does X have? (1 point)

    _____

    c. Which element in its Group could X be if it had the greatest density? (1 point)

    _____

    d. Name another element besides Cl that would have the same subscripts in a compound with X. (1 point)

    _____

5. If the strength of attraction of an atom for its outermost electrons determines the size of an atom

    a. What three conditions of an atom control its strength of attraction and size? (3 points)

    b. Explain why, as the atomic number increases in a *Group* on the Periodic Table of the Elements, the size of atoms *increases*. (1 point)

    c. Explain why, as the atomic number increases in a *Period* on the Periodic Table of the Elements, the size of atoms *decreases*. (1 point)

CHAPTER 3
MOLES/STOICHIOMETRY

Pure substances can be classified as either *elements* or *compounds*.

Elements are the simplest kinds of substances. They are made up of atoms that are essentially all alike and have the same atomic number. They cannot be broken down into simpler substances by chemical change. Carbon, hydrogen, oxygen, iron, and sulfur are examples of elements.

Compounds are substances made up of two or more elements that are chemically combined in a fixed proportion. These substances *can* be broken into their component elements by chemical change. The properties of compounds are generally quite different from those of the elements of which they are composed. H_2O, CO_2 and NaCl are examples of compounds.

CHEMICAL FORMULAS

Chemical formulas use symbols and numerals to provide both qualitative and quantitative information about the composition of substances.

Symbols

Chemical symbols are abbreviations for the elements. Symbols consist of one or two letters. A single-letter symbol is always a capital, or upper-case, letter. In a two-letter symbol, the first letter is always a capital letter. The second letter is always a small, or lower-case, letter.

Formulas

A formula is actually a statement informing us of the identity of the elements present in a substance and the atomic ratios in which they are found in such substance.

Empirical formulas. The formula that depicts the *simplest* atomic ratio in which the elements can combine in a compound is the **empirical formula** of that compound. Formulas for ionic compounds are usually empirical formulas. For example, sodium chloride consists of 1 ion of sodium (Na) for each ion of chlorine (Cl). Its formula is NaCl. Magnesium chloride consists of two ions of chlorine for each ion of magnesium (Mg). Its formula is $MgCl_2$.

Molecular formulas. Covalent structures are comprised of molecules. These molecules sometimes consist of atoms combined chemically in *multiples* of their simplest ratio. In such cases, the empirical formula, while useful, does not tell the whole story. For example, there are many compounds made up of the two elements carbon and hydrogen. The empirical formula for one group of these compounds is CH_2. The ratio of carbon atoms to hydrogen atoms is 1 to 2. Laboratory analysis of a compound from this group may reveal that one molecule of the compound has a mass that is four times the combined masses of one carbon and two hydrogen atoms. Thus, the molecular formula for this particular compound must be 4 times CH_2, or C_4H_8.

NAMING AND WRITING FORMULAS OF CHEMICAL COMPOUNDS

In most cases, chemical formulas are determined by laboratory procedures. Names of compounds are based on the formulas.

Binary compounds are composed of two elements. In binary compounds composed of a metal and a nonmetal, the metallic element is named and written first. The name of such compounds ends in *-ide*. Most compounds of this type are binary *salts*. For example, the compound made up of calcium (metal) and chlorine (nonmetal) has a formula of $CaCl_2$ and is named calcium chloride.

In binary compounds composed of two nonmetals, the less electronegative element is named and written first. The name of the compound ends in *-ide*. Prefixes are used to indicate the number of atoms of each nonmetal. For example, consider two compounds made up of carbon and oxygen. One, having one atom of each element, has the formula CO and is named carbon *mon*oxide. The other, having one carbon atom and two oxygen atoms, has the formula CO_2 and is named carbon *di*oxide. Some other prefixes used include *tri-* (3), as in sulfur trioxide (SO_3), and *tetra-* (4), as in carbon *tetra*chloride (CCl_4).

Binary acids consist of hydrogen plus some nonmetal. The rule for naming these acids is:

prefix *hydro-* + stem name of nonmetal + suffix *-ic*

Examples:
In HCl, the nonmetal is *chlor*ine, the name of the acid is hydrochloric acid.
In HBr, the nonmetal is *brom*ine, the name of the acid is hydrobromic acid.

Ternary compounds contain three different kinds of atoms. Most compounds that contain three or more different kinds of atoms contain polyatomic ions. Bases consist of metallic ions combined with the hydroxide ion, OH^-. The names of the bases consist of the names of the metallic ion followed by the word *hydroxide*.

Examples:
NaOH is named sodium hydroxide.
$Mg(OH)_2$ is named magnesium hydroxide.

Ternary acids consist of hydrogen ions combined with polyatomic ions. Names of ternary acids are based on the names of the polyatomic ions. Prefixes and suffixes are applied as needed. If the name of the polyatomic ion ends in *-ite*, the name of the acid ends in *-ous*. If the name of the polyatomic ion ends in *-ate*, the name of the acid ends in *-ic*. (*Compare Reference Tables for Physical Setting/Chemistry* Tables L and K with Table E).

Examples:

Polyatomic ion	Acid	Acid name
-SO_4 (sulf*ate*)	H_2SO_4	sulfur*ic* acid
-SO_3 (sulf*ite*)	H_2SO_3	sulfur*ous* acid

Stock System. In naming compounds of metals that have more than one oxidation number, the Stock system is used. In this system, the name of the metal is followed by a Roman numeral that represents its oxidation number in that compound. The rest of the name is determined by the appropriate rule.

Examples:

FeO is iron (II) oxide

Fe_2O_3 is iron (III) oxide

$CuSO_4$ is copper (II) sulfate

Cu_2SO_4 is copper (I) sulfate

To write a correct formula, the oxidation numbers of the elements must be known. When elements are ions, their oxidation number is the same as their charge. Elements in the same group usually have the same charge.

Group	Charge	Example
1	+1	Na^+
2	+2	Ba^{2+}
13	+3	Al^{3+}
15	-3	N^{3-}
16	-2	O^{2-}
17	-1	Br^-

When more than one positive oxidation number is possible (as for many elements in groups 3-11) the charge will be indicated in the name of the element as a Roman numeral. For example: iron in iron III oxide is +3. For polyatomic ions, the *Reference Tables for Physical Setting/Chemistry* Table E gives the oxidation number. The positive element comes first in the name and in the formula.

For all neutral formulas, the oxidation numbers must add up to zero. In the case of sodium sulfide, Na^+ and S^{2-}, would need two Na^+ and one S^{2-} to have no net charge. Its formula would be Na_2S, with the number of atoms as subscripts (although the number 1 is understood and not written).

Below are the names of some compounds, their ions, and their formulas:

Compound	Ions	Formula
sodium oxide	Na^+ and O^{2-}	Na_2O
calcium chloride	Ca^{2+} and Cl^-	$CaCl_2$
potassium sulfate	K^+ and SO_4^{2-}	K_2SO_4
aluminum nitrate	Al^{3+} and NO_3^-	$Al(NO_3)_3$
barium phosphate	Ba^{2+} and PO_4^{3-}	$Ba_3(PO_4)_2$
chromium III bromide	Cr^{3+} and Br^-	$CrBr_3$
nitrogen IV oxide	N^{+4} and O^{2-}	NO_2

MOLE INTERPRETATION

Since finding the mass of a single atom would be virtually impossible, some standard unit is needed to compare conveniently the masses in grams of the atoms of elements or compounds. The term mole is now used to name this number in the same way a dozen is used to name the number 12 or a gross is used to name the number 144.

The term **mole** was selected to indicate the number of molecules in a *mole*cular mass of a compound, expressed in grams, or the number of atoms in the atomic mass of an element, expressed in grams. The number has since been found to be 6.02×10^{23}. It has been named *Avogadro's number* in honor of the Italian professor of physics.

USE OF THE MOLE CONCEPT

Gram atomic mass (gram-atom)

The mass of 6.02×10^{23} atoms, or 1 mole of atoms, of an element represents the **gram atomic mass** of that element. It is also called one **gram-atom** of the element. This quantity is numerically equal to the atomic mass of the element, which can be found in the periodic table.

PROBLEMS INVOLVING FORMULAS

Gram Molecular Mass

The mass of 6.02×10^{23} molecules, or 1 mole of molecules, of a substance represents the **gram molecular mass,** or **molar mass** of that substance. It is found by adding the atomic masses of all the atoms in a given molecule of the substance.

The sum of the masses of ions, or one mole of empirical units, of an ionic substance is called the **formula mass** of that substance. This mass is used in calculations dealing with ionic substances and also network solids, since they are not composed of molecules.

SAMPLE PROBLEM

Find the gram formula mass for sodium sulfate.

Solution:

Step 1. Write the correct formula Na_2SO_4
Step 2. Find the gram atomic mass of each element in the periodic table.
Step 3. Multiply each atomic mass by the subscript in the formula.
Step 4. Find the sum of all the masses.

Element	g-atomic Mass		Subscript		Totals
Na	23g	×	2	=	46g of Na
S	32g	×	1	=	32g of S
O	16g	×	4	=	64g of O
One mole of Na_2SO_4				=	**142g of Na_2SO_4**

Find the mass of one mole of glucose, $C_6H_{12}O_6$.

$$C = \quad 12g \times \quad 6 \ = 72g\ C$$
$$H = \quad 1g \times \quad 12 = 12g\ H$$
$$O = \quad 16g \times \quad 6 \ = 96g\ O$$

Answer = $\underline{\qquad 180g\ C_6H_{12}O_6 \qquad}$

SAMPLE PROBLEM

What is the mass of 4.00 moles of sodium sulfide?

Step 1. Determine the formula of sodium sulfide:
$Na^+ + S^{2-}$ gives Na_2S

Step 2. Find the gram formula mass from the Periodic Table:
two Na: 2 x 23.0 g/mol   =   46.0 g/mol
one S: <u>32.0 g/mol</u>
                               78.0 g/mol

Step 3. Multiply the mass of one mole (gram formula mass) by 4.00 moles:
78.0 g/mol x 4.00 mol = 312 grams

SAMPLE PROBLEM

How many moles are 25.0 grams of calcium hydroxide?

Solution:

Step 1. Determine the formula of calcium hydroxide
Ca^{2+} + OH^- gives $Ca(OH)_2$

Step 2. Find the gram formula mass from the Periodic Table:
one Ca :   40.1 g/mol
two H :   2 x 1.0 g/mol = 2.0 g/mol
two O :   2 x 16.0 g/mol = 32.0 g/mol
                              74.1 g/mol

Step 3. Divide the mass by the mass of one mole (gram formula mass):

$$\frac{25.0g}{74.1 \ ^{g/mol}} = \textbf{0.337 moles}$$

Percent Composition. Since the formula of a compound provides the mole ratios of its component atoms, it is possible to determine the number of moles of each element present in a given mass of the compound. With this information, the percentage, by mass, of each element can then be calculated.

SAMPLE PROBLEM

Determine the percent composition of mercuric (II) chloride, $HgCl_2$.

Solution:

Step 1. Calculate the formula mass of $HgCl_2$:
$$201 + (2 \times 35.5) = 272$$

Step 2. Calculate the percentage of the formula mass due to each component:

$$\frac{\text{mass due to Hg}}{\text{formula mass}} = \frac{201}{272} = 0.739 \text{ or } \textbf{73.9\% Hg}$$

$$\frac{\text{mass due to Cl}_2}{\text{formula mass}} = \frac{71.0}{272} = 0.261 \text{ or } \textbf{26.1\% Cl}$$

Check: 73.9% + 26.1% = 100.0%

Empirical Formula. The simplest mole ratio among the component elements in a compound is represented by whole-number subscripts in the empirical formula for that compound. The molecular formula for the compound can be any simple multiple of the empirical formula, depending on the molar mass of the compound.

Consider the group of compounds made up of atoms of carbon and hydrogen. For the compound that contains 2 moles of hydrogen for each mole of carbon, the empirical formula is CH_2. The empirical formula mass of this compound is:

$$
\begin{aligned}
C &= 12g \quad \times\ 1\ = \quad 12g\ carbon \\
H &= 1g \quad\ \times\ 2\ = \quad \underline{2g\ hydrogen} \\
&\qquad\qquad\qquad\qquad\ 14g\ CH_2
\end{aligned}
$$

If laboratory analysis reveals that the gram molecular mass of a compound is actually 14g, then CH_2 is the molecular, as well as the empirical formula of the compound. However, suppose the gram molecular mass is found to be 42g/mole. Since $42 \div 14 = 3$, the molecular formula for the compound must be $3(CH_2)$, or C_3H_6.

SAMPLE PROBLEM

Empirical Formula from Percent Composition

Find the empirical formula of a compound composed of 75% carbon and 25% hydrogen by mass.

Solution:

Step 1.   Assume a 100-gram sample.

Step 2.   Find the mass of each element in the sample:
mass of C   = 75% of 100 grams = 75g C
mass of H   = 25% of 100 grams = 25g H

Step 3.   Convert grams to moles:

$$75g\ C\ \times\ \frac{1\ mole\ C}{12g\ C} = 6.3\ moles\ C$$

$$25g\ H\ \times\ \frac{1\ mole\ H}{1.0g\ H} = 25\ moles\ H$$

Step 4.   Find the mole ratio (divide both numbers by the smaller number):

$6.3 \div\ 6.3 = 1.0\ C$

$25\ \ \div\ 6.3 = 4.0\ H$

empirical formula = CH_4

SAMPLE PROBLEM

Molecular Formula from Empirical Formula and Gram Molecular Mass

A compound with empirical formula CH_3 is found to have a gram molecular mass of 30. grams. Find its molecular formula.

Solution:

Step 1.   Find the empirical formula mass of CH_3:

$$C = 12g \times 1 = 12g\ C$$
$$H = \ \ 1g \times 3 = \ \underline{3g\ H}$$
$$15g\ = \text{empirical formula mass of } CH_3$$

Step 2.   Find the molecular formula:

$$\text{molecular formula} = \frac{\text{empirical}}{\text{formula}} \times \frac{\text{molecular mass}}{\text{empirical formula mass}}$$

$$= CH_3 \times \frac{30.g}{15g}$$

$$= CH_3 \times 2\ = C_2H_6$$

$$\text{molecular formula} = \mathbf{C_2H_6}$$

QUESTIONS

1. Which of the following substances can *not* be decomposed by chemical change?

   (1) sulfuric acid   (2) ammonia   (3) water   (4) argon

2. Which substance is composed of atoms that all have the same atomic number?

   (1) magnesium   (2) methane   (3) ethane   (4) ethene

3. Which is a binary compound?

   (1) potassium hydroxide   (2) magnesium sulfate
   (3) aluminum oxide   (4) ammonium chloride

4. The molecular mass of a compound of carbon and hydrogen is 42. Its empirical formula is

    (1) CH (2) CH_2 (3) CH_3 (4) CH_4

5. Which formula is an empirical formula?

    (1) H_2CO_3 (2) $H_2C_2O_4$ (3) CH_3COOH (4) CH_2OHCH_2OH

6. Which substance has the same molecular and empirical formulas?

    (1) C_6H_6 (2) C_2H_4 (3) CH_4 (4) $C_6H_{12}O_6$

7. The approximate percent by mass of potassium in $KHCO_3$ is

    (1) 19%   (2) 24%   (3) 39%   (4) 61%

8. The percent by mass of nitrogen in NH_4NO_3 (formula mass=80) is approximately

    (1) 18%   (2) 23%   (3) 32%   (4) 35%

9. The percent by mass of nitrogen in $Mg(CN)_2$ is equal to

    (1) $\frac{14}{76} \times 100$ (2) $\frac{14}{50} \times 100$ (3) $\frac{28}{76} \times 100$ (4) $\frac{28}{50} \times 100$

10. Which is the correct formula for titanium (III) oxide?

    (1) Ti_2O_3 (2) TiO (3) Ti_3O_2 (4) Ti_2O_4

11. Which is the correct formula for nitrogen (II) oxide?

    (1) NO (2) NO_2 (3) NO_3 (4) NO_4

12. What is the name for the sodium salt of the acid $HClO_2$?

    (1) sodium chlorite   (2) sodium chloride
    (3) sodium chlorate   (4) sodium perchlorate

13. What is the correct formula for sodium sulfate?

    (1) $Na_2S_2O_4$ (2) Na_2SO_3 (3) Na_2SO_4 (4) $Na_2S_2O_3$

14. Which is the formula for magnesium sulfide?

(1) MgS (2) $MgSO_3$ (3) MnS (4) $MnSO_3$

15. A compound has an empirical formula of CH_2 and a molecular mass of 56. Its molecular formula is

(1) C_2H_4 (2) C_3H_6 (3) C_4H_8 (4) C_5H_{10}

16. What is the correct formula for chromium (III) oxide?

(1) CrO_3 (2) Cr_3O (3) Cr_2O_3 (4) Cr_3O_2

17. What is the molecular formula of a compound that has a molecular mass of 92 and an empirical formula of NO_2?

(1) NO_2 (2) N_2O_4 (3) N_3O_6 (4) N_4O_8

18. What is the correct formula of potassium hydride?

(1) KH (2) KH_2 (3) KOH (4) $K(OH)_2$

19. Element X forms the compounds XCl_3 and X_2O_3. In the Periodic table, element X would most likely be found in Group

(1) 1   (2) 2   (3) 13   (4) 14

20. What is the formula mass of $Al_2(SO_4)_3$?

(1) 123   (2) 150   (3) 214   (4) 342

21. Which represents the greatest mass of chlorine?

(1) 1 mole of chlorine   (2) 1 atom of chlorine
(3) 1 gram of chlorine   (4) 1 molecule of chlorine

22. What is the total number of atoms of oxygen in the formula $Al(ClO_3)_3 \cdot 6H_2O$?

(1) 6   (2) 9   (3) 10   (4) 15

23. What is the total number of moles of atoms present in 1 mole of $Ca_3(PO_4)_2$?

(1) 13   (2) 10   (3) 8   (4) 5

24. In the compound Al_2O_3, the ratio of aluminum to oxygen is

(1) 2 grams of aluminum to 3 grams of oxygen
(2) 3 grams of aluminum to 2 grams of oxygen
(3) 2 moles of aluminum to 3 moles of oxygen
(4) 3 moles of aluminum to 2 moles of oxygen

25. What is the mass in grams of 2.00 moles of $CaCl_2$?

(1) 1.00   (2) 2.00   (3) 111   (4) 222

26. How many moles are 29 grams of NaCl?

(1) 0.50   (2) 2.0   (3) 1.0   (4) 58

27. In which Group do the elements usually form oxides which have the general formula M_2O_3?

(1) 1   (2) 2   (3) 13   (4) 14

CHEMICAL EQUATIONS

A chemical equation usually represents a chemical reaction. An equation should identify:

1. the reactants and products.

2. the mole ratios of these species.

3. some reference to the energy changes involved.

4. phases of reactants and products.

When writing chemical equations, the convention is to use an arrow to indicate the direction of the reaction. (Double arrows are reversible reactions). Reactants are written to the left of the arrow, products to the right of the arrow. Energy is indicated in various ways. (See Page 91)

It is important that equations conform to the Law of Conservation of Mass. This means that the total number of atoms of each element present among the products must be the same as that found among the reactants.

In the equation:

$$2H_2(g) + O_2(g) \rightarrow 2H_2O\ (\ell)\ + heat$$

the following information is given:

- 4 atoms of hydrogen gas react with 2 atoms of oxygen gas.
- 2 molecules of liquid water are produced and heat is released.
- 2 moles of hydrogen gas, each consisting of 4 moles of hydrogen atoms, will combine with 1 mole of oxygen gas, consisting of 2 moles of oxygen atoms, to produce

2 moles of liquid water, each consisting of 4 moles of hydrogen atoms and 2 moles of oxygen atoms. Heat energy is also produced by this reaction.

The ratio of $H_2(g)$ molecules reacting to $O_2(g)$ molecules is 2 to 1 or 2/1. It is also the ratio that *moles* of $H_2(g)$ react to *moles* of $O_2(g)$. this 2/1 ratio is known as the **mole ratio** and is essential in comparing amounts that react or are produced.

STOICHIOMETRY

Calculations based on quantitative relationships in a *balanced chemical equation* are referred to as **stoichiometry**. This term comes from the Greek words *stoicheon* (element) and *metrein* (to measure).

In order to use stoichiometric methods, we must assume that:

1. the reaction has no side reactions.
2. the reaction goes to completion.
3. a reactant is completely consumed.

Problems Involving Equations

A balanced equation provides a considerable amount of information about the reaction it represents. It not only identifies the reactant and products, it also gives the mole proportions of all the substances involved in the reaction. These proportions are shown by the coefficients used to balance the equation. Consider the following:

$$2Mg(s) + O_2(g) \rightarrow 2MgO(s)$$

This equation tells us that 2 moles of Mg metal react with 1 mole of oxygen gas to produce 2 moles of magnesium oxide (a white powder). Using this information, it is possible to calculate the moles of any substance involved in this reaction given the moles of any one of the other substances involved.

SAMPLE PROBLEM

How many moles of magnesium are needed to produce 3.0 moles of magnesium oxide in the reaction: $2Mg(s) + O_2(g) \rightarrow 2MgO(s)$

Solution:

Step 1. Identify which substance is known and which substance is unknown.
Known = 3.0 mol of MgO
Unknown = x mol of Mg

Step 2. Find the mole ratio from the coefficients of the balanced equation

$$\frac{2 \text{ mol Mg}}{2 \text{ mol MgO}}$$

note: The unknown is always in the numerator of the mole ratio.

Step 3. Multiply the known moles by the mole ratio.

$$3.0 \text{ mol MgO} \times \frac{2 \text{ mol Mg}}{2 \text{ mol MgO}} = \textbf{3.0 mol Mg}$$

SAMPLE PROBLEM

How many moles of ammonia will be produced when 4.0 moles of hydrogen reacts with sufficient nitrogen in the reaction:

$$N_2(g) + 3H_2(g) \rightarrow 2NH_3(g)$$

Solution:

Step 1. Identify which substance you know and which substance is unknown.
Known = 4.0 mol of H_2
Unknown = X mol of NH_3

Step 2. Find the mole ratio from the coefficients of the balanced equation.

$$\frac{2 \text{ mol NH}_3}{3 \text{ mol H}_2}$$

Step 3. Multiply the known moles by the mole ratio.

$$4.0 \text{ mol H}_2 \times \frac{2 \text{ mol NH}_3}{3 \text{ mol H}_2} = \textbf{2.7 mol NH}_3$$

CHAPTER 3

QUESTIONS

1. When the equation $\_\_Al(s) + \_\_O_2(g) \rightarrow \_\_Al_2O_3(s)$ is correctly balanced using the smallest whole numbers, the coefficient of Al(s) is

   (1) 1　　　　(2) 2　　　　(3) 3　　　　(4) 4

2. Given the unbalanced equation

   $$\_\_Al_2(SO_4)_3 + \_\_Ca(OH)_2 \rightarrow \_\_Al(OH)_3 + \_\_CaSO_4$$

   When the equation is completely balanced using the smallest whole-number coefficients, the sum of the coefficients is

   (1) 5　　　　(2) 9　　　　(3) 3　　　　(4) 4

3. Given the equation: $\_\_FeCl_2 + \_\_Na_2CO_3 \rightarrow \_\_FeCO_3 + \_\_NaCl$
   When the equation is correctly balanced using the smallest whole numbers, the coefficient of NaCl is

   (1) 6　　　　(2) 2　　　　(3) 3　　　　(4) 4

4. Given the balanced equation: $2Na + 2H_2O \rightarrow 2X + H_2$
   What is the correct formula for the product represented by the letter X?

   (1) NaO　　　　(2) Na_2O　　　　(3) NaOH　　　　(4) Na_2OH

5. Given the balanced equation:

   $$3Fe + 4H_2O \rightarrow Fe_3O_4 + 4H_2$$

   What is the mole ratio of H_2 produced to Fe reacting?

   (1) ½　　　　(2) ⁴/₃　　　　(3) ³/₄　　　　(4) ¼

6. 5.0 grams of X reacts with 8.0 grams of Y to produce Z and 3.0 grams of Q according to the equation: $2X + Y \rightarrow 2Z + 3Q$.
   How many grams of Z are produced?

   (1) 10　　　　(2) 2.0　　　　(3) 3.0　　　　(4) 13

7. Given the reaction:

$$(NH_4)_2CO_3 \rightarrow 2NH_3 + CO_2 + H_2O$$

What is the minimum amount of ammonium carbonate that reacts to produce 1.0 mole of ammonia?

(1) 0.25 mole   (2) 0.50 mole   (3) 17 moles   (4) 34 moles

8. Given the reaction:

$$4Na + O_2 \rightarrow 2Na_2O$$

How many moles of oxygen are completely consumed in production of 1.00 mole of Na_2O?

(1) 1.0   (2) 2.0   (3) 1.5   (4) 0.50

9. Given the reaction:

$$4Al + 3O_2 \rightarrow 2Al_2O_3$$

How many moles of Al_2O_3 will be formed when 2.0 moles of Al reacts completely with O_2?

(1) 1.0   (2) 2.0   (3) 0.50   (4) 4.0

10. Given the reaction:

$$C_3H_8(g) + 5O_2(g) \rightarrow 3CO_2(g) + 4H_2O(g)$$

How many moles of $H_2O(g)$ formed when 3.0 moles of $C_3H_8(g)$ is completely oxidized?

(1) 0.75   (2) 12   (3) 3.0   (4) 4.0

CHAPTER 3

TYPES OF CHEMICAL REACTIONS

Chemical equations are used to describe chemical reactions. There are several types of chemical reactions. Some basic types are listed below.

1. *Synthesis* is a reaction in which a compound is made from two or more elements or simpler compounds. An example is:

$$N_2(g) + 2\,O_2(g) \rightarrow 2\,NO_2(g)$$

2. *Decomposition* or (analysis) is the opposite of synthesis. In decomposition a compound is broken down into its elements or simpler compounds. An example is:

$$2\,H_2O\,(\ell) \rightarrow 2\,H_2(g) + O_2(g)$$

3. *Single replacement* is a reaction in which an element replaces another element in a compound. This type of reaction is always an oxidation-reduction reaction (see Chapter 8). An example is:

$$Cu\,(s) + 2\,AgNO_3(aq) \rightarrow Cu\,(NO_3)_2\,(aq) + 2\,Ag\,(s)$$

4. *Double replacement* reactions are ones in which two compounds react, and two new compounds are formed. The compounds exchange partners. A product of double replacement is a precipitate or water. Acid-base reactions (see Chapter 9) and precipitation reactions are usually double replacement. An example is:

$$Na_2CO_3(aq) + Ca(NO_3)_2\,(aq) \rightarrow CaCO_3(s) + 2\,NaNO_3(aq)$$

PRACTICE FOR CONSTRUCTED RESPONSE

1. Below are four reactions,

(1) _____ $H_2SO_4(aq)$ + _____ $NaOH(aq)$ → _____ $Na_2SO_4(aq)$ + _____ $H_2O(\ell)$

(2) _____ $N_2O_5(g)$ → _____ $N_2(g)$ + _____ $O_2(g)$

(3) _____ $N_2(g)$ + _____ $H_2(g)$ → _____ $NH_3(g)$

(4) _____ $Al(s)$ + _____ $Fe(NO_3)_2(aq)$ → _____ $Al(NO_3)_3(aq)$ + _____ $Fe(s)$

   a. Balance the four equations with the smallest whole numbers. (4 points)

   b. Which reaction is single replacement? _____
(1 point)

   c. Which reaction is double replacement? _____
(1 point)

   d. Which reaction is synthesis? _____
(1 point)

   e. Which reaction is decomposition? _____
(1 point)

2. If iron ore is Fe_2O_3

   a. Find its formula mass. (1 point)

_____

   b. Calculate the percent by mass of the ore that is iron. (2 points)

_____

   c. Calculate the mass of iron that can be obtained from 500. grams of iron ore. Show your work here. (2 points)

_____

_____

3. A hydrocarbon in gasoline is analyzed and found to contain 14.3%H. The rest is carbon.

    a. Calculate the empirical formula. Show your work here. (2 points)

_____

    b. If the molecular mass is found to be 84.0, determine the molecular formula. (1 point)

_____

4. The equation for the complete combustion of propane is

$$C_3H_8(g) + 5\,O_2(g) \rightarrow 3\,CO_2(g) + 4\,H_2O(g)$$

    a. If 10. molecules of oxygen react, how many molecules of water are produced? Show all work here. (2 points)

_____

    b. If 3.0 moles of C_3H_8 react, how many moles of CO_2 are produced? Show all work here. (2 points)

_____

    c. If 11 g of C_3H_8 reacts completely with 40. g of O_2 producing 18 g. of H_2O, how many grams of CO_2 are produced? Show all work here. (2 points)

_____

5. Using the structural formulas:
H - H for an H_2 molecule
O - O for an O_2 molecule

O ⟨ H for an H_2O molecule
    H

Show how many H_2O molecules will be produced and how many oxygen molecules will be left unreacted if 4 hydrogen molecules react as much as possible with 5 oxygen molecules. Use the boxes below to draw your structural formulas. (3 points)

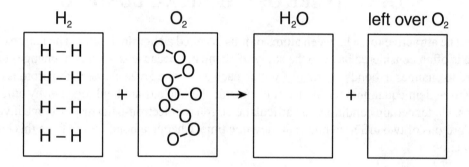

6. For 13.2 g of Ca (OH)$_2$

  a. Calculate the number of moles. _____
     (2 points)

  b. How many significant figures must your answer have? _____
     (1 point)

CHAPTER 4
CHEMICAL BONDING

THE NATURE OF CHEMICAL BONDING

The attractive force between atoms or ions is called a **chemical bond.** The nature of this force is often considered basic to the study of chemistry because all chemical changes can be traced to changes in bonds. Normally, within each atom of a given element, the protons of the nucleus and an equal number of electrons outside the nucleus are held together by attractive forces. Under certain conditions, a particular electron (or electron-pair) may become involved with protons of two different nuclei at the same time. Chemical bonds result from this kind of interaction.

Chemical Energy

Energy based on position is called potential energy. (See page 62) The particular form of potential energy involved in the making and breaking of chemical bonds is called **chemical energy.**

Energy Changes in Bonding

Whenever chemical bonds are formed, energy is released (an exothermic process); when bonds are broken, energy is absorbed (an endothermic process). It follows then, that when two atoms are bonded to each other, they are at a lower energy level than when they are alone, because the total energy is conserved. If energy is lost, the remaining potential energy must be lower.

Bonding and Stability

Stability is inversely related to potential energy. In other words, systems at lower energy levels are said to be more stable. Whenever large amounts of energy are released in the formation of a bond, the bond is said to be strong and the system very stable. Weak bonds and unstable systems are associated with the release of small amounts of energy. This relationship is very important in the study of kinetics and thermochemistry (Heat of Reaction, Heat of Formation, etc.).

Electronegativity

Electronegativity is the name given to an atom's ability to attract electrons to itself in bond formation. Electronegativity values cannot be measured directly. They are obtained from a variety of data and from a variety of assumptions. The scale of electronegativity values is an arbitrary one. The highest value on the scale is 4.0, assigned to fluorine, the most electronegative of the elements.

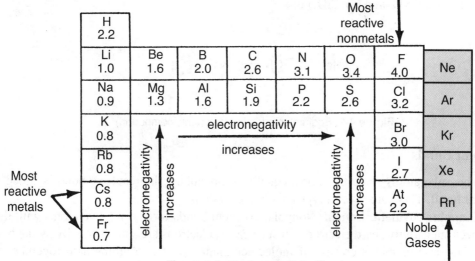

Figure 4-1. Electronegativity Chart

THE BONDS BETWEEN ATOMS

Chemical bonds form when two atoms share electrons or when electrons are transferred from one atom to another. In most cases, the electrons involved in the bonding process are valence electrons. As a result of bond formation, both atoms achieve a valence shell that is identical to one of the noble (inert) gases. This condition—8 valence electrons—is associated with maximum stability and minimum potential energy content. It is known as the **stable octet**.

Ionic Bonds

Whenever electrons are transferred from one atom to another, the atom that loses electrons becomes a positively charged ion and its radius decreases. The atom receiving additional electrons becomes a negatively charged ion and its radius increases. The force of attraction that holds the oppositely charged ions together is called an **ionic bond.**

The ions produced as a result of electron transfer usually have electron configurations identical to those of noble gases. The resulting compounds are said to be ionic. In the solid phase, ionic compounds are held together in very rigid, fixed crystalline structures. Ionic solids are poor conductors of electricity and have high melting points. When ionic solids are melted or dissolved in water, the ions become free to move about and are therefore capable of conducting electricity.

Salts are the most common examples of ionically bonded substances. They can be defined as a metal bonded to a nonmetal (binary salts) or a metal bonded to a polyatomic ion (ternary salts). The difference in electronegativity between the bonding atoms must be very large - usually 1.7 or greater. Examples of binary salts are: $NaCl$, MgS or $FeBr_2$. Examples of ternary salts are KNO_3 and $Ca(ClO_3)_2$.

$$Na^x \;+\; \cdot\overset{\displaystyle\cdot\cdot}{\underset{\displaystyle\cdot\cdot}{C}}l\!: \;\rightarrow\; Na^+ \left[\overset{\displaystyle\cdot\cdot}{\underset{\displaystyle\cdot\cdot}{x\,C\,l}}\!:\right]^-$$

Figure 4-2. Sodium chloride is a salt (ionic compound)

Covalent Bonds

When two atoms share electrons, equally or unequally, the resulting bond is called a **covalent bond.** There are several types of covalent bonds.

Nonpolar covalent bonds. Nonpolar covalent bonds form when two atoms with the same electronegativity share electrons. In such cases, electrons are shared equally by the two atoms. Diatomic elements consist of molecules made up of two atoms held together by nonpolar covalent bonds. The chlorine molecule (Cl_2) is one example.

$$\overset{\displaystyle xx}{\underset{\displaystyle xx}{\!^x_xCl}}\!:\!\overset{\displaystyle\cdot\cdot}{\underset{\displaystyle\cdot\cdot}{C}}l\!:$$

Figure 4-3 The Cl_2 molecule has a nonpolar covalent bond.

Polar covalent bonds. Polar covalent bonds form when atoms with different electronegativities share electrons. The polarity is caused by the fact that the electrons are not shared equally. In the hydrogen chloride molecule (HCl), the bond holding the hydrogen and chlorine atoms together is a polar covalent bond.

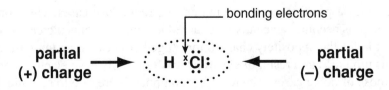

partial (+) charge → H ^{x}Cl: ← **partial (−) charge**

bonding electrons

Figure 4-4. The HCl molecule has a polar covalent bond.

Symmetry and Polarity

The asymmetry of the charge in a molecule creates the polarity. In bonding, the eight valence electrons are arranged in pairs around an atom. Because of the sameness of their charges, the pairs repel each other to their farthest positions from each other. This puts the electron pairs at the four corners of a geometric shape known as a tetrahedron, with the nucleus of the atom at the center.

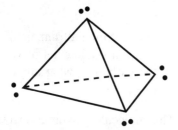

Notice that the figure is not flat. The four sides are equilateral triangles.

Figure 4-5. Tetrahedron

Directional Nature of Covalent Bonds

Many properties of compounds can be related to the shapes of their molecules. These shapes, in general, are the result of the directional nature of covalent bonds. A good example is the polarity of the water molecule. The molecule is polar because of its bent shape. This asymmetrical shape causes the non-uniform distribution of charge, which results in the polarity.

For water, H_2O: The oxygen has six valence electrons, and the hydrogen have one each. The hydrogens share their electrons with the unpaired electrons of oxygen at two of the corners of the tetrahedron. This predicts an asymmetric *bent shape* for the molecule

There is a large difference in electronegativity causing the oxygen end to attract electrons more. It is also asymmetric so the molecule is polar.

For ammonia, NH_3: The nitrogen has five valence electrons, one pair and three unpaired. The hydrogens bond with the unpaired nitrogen electrons at three corners of the tetrahedron. This predicts the asymmetric *pyramidal shape* for the molecule.

There is a large difference in electronegativity causing the nitrogen end to attract the electrons more. The molecule is also asymmentric so it is polar.

For methane, CH_4: The carbon has four unpaired valence electrons. The hydrogens share electrons with each of these carbon electrons at all four corners of the *tetrahedron*.

Because this molecule is symmetrical it will not be polar regardless of the electronegativities.

For Carbon dioxide, CO_2: the carbon has four valence electrons, all unpaired, and each oxygen has six valence electrons, two unpaired. To achieve a stable octet of valence electrons, each oxygen shares two pairs of electrons with the carbon in what are known as **double covalent** bonds. The resulting molecule has a *linear shape*.

This molecule is symmetrical so it is nonpolar regardless of its electronegativities.

Molecular substances. Molecules are particles that consist of covalently bonded atoms, such as H_2, NH_3, H_2O, HCl, CCl_4, and S_8 to name a very few.

Molecular substances may exist as gases, liquids, or solids, depending on the degree of attraction that exists among the molecules. Molecular solids are usually soft, poor conductors of heat and electricity, and have low melting points.

Polyatomic Ions

Most polyatomic ions:
1. are held together by covalent bonds
2. have charges making them ionic

Example: S - O Bonds are covalent, but the SO_4 group has a total charge of 2–. NH_4^+, CO_3^{1-}, ClO_3^{1-}, PO_4^{3-} are other examples.

Metallic bonds. Metals usually have one , two, or three valence electrons. These electrons are readily lost or transferred and, in a metallic solid, they are not always associated with any particular atom. Therefore, the particles in a metal are actually positive ions surrounded by very mobile electrons that can "drift" from one atom to another ("sea" of mobile electrons). The strong bonds that result from the attraction of all the positive ions for the electrons surrounding them are called **metallic bonds**. Their strength accounts for the high melting points of metals. The mobile electrons account for the luster of metals and their ability to conduct heat and electricity.

QUESTIONS

1. Which kind of energy is stored within a chemical substance?

   (1) free energy   (2) activation energy
   (3) kinetic energy   (4) potential energy

2. Which type of bonding is characteristic of a substance that has a high melting point and electrical conductivity only in the liquid phase?

   (1) nonpolar covalent   (2) polar covalent
   (3) ionic   (4) metallic

3. Which diagram best represents a polar molecule?

   Cl_2 H_2 HCl   NaCl

   (1)   (2)   (3)   (4)

4. Atoms of which of the following elements have the strongest attraction for electrons?

   (1) aluminum   (2) chlorine   (3) silicon   (4) sodium

5. Which element is most likely to gain electrons in a chemical reaction?

   (1) Kr   (2) Br   (3) Ca   (4) Ba

6. Which substance exists as a metallic crystal at STP?

   (1) Ar   (2) Au   (3) SiO_2 (4) CO_2

7. The P—Cl bond in a molecule of PCl_3 is

   (1) nonpolar   (2) polar covalent
   (3) metallic   (4) electrovalent

8. The electrons in a bond between two iodine atoms (I_2) are shared

   (1) equally, and the resulting bond is polar
   (2) equally, and the resulting bond is nonpolar
   (3) unequally, and the resulting bond is polar
   (4) unequally, and the resulting bond is nonpolar

9. What type of bond exists in a molecule of hydrogen iodide?

   (1) a polar covalent bond with an electronegativity difference of zero
   (2) a polar covalent bond with an electronegativity difference between zero and 1.7
   (3) a nonpolar covalent bond with an electronegativity difference of zero
   (4) a nonpolar covalent bond with an electronegativity difference between zero and 1.7

10. Given the reaction:

$$H_2 + Cl_2 \rightarrow 2HCl$$

    Which statement best describes the energy change as bonds are formed and broken in this reaction?

    (1) The breaking of the Cl—Cl bond releases energy.
    (2) The breaking of the H—H bond releases energy.
    (3) The forming of the H—Cl bond absorbs energy.
    (4) The forming of the H—Cl bond releases energy.

11. Which atoms are most likely to form covalent bonds?

    (1) metal atoms that share electrons
    (2) metal atoms that share protons
    (3) nonmetal atoms that share electrons
    (4) nonmetal atoms that share protons

12. Which formula represents an ionic compound?

    (1) KCl
    (2) HCl
    (3) CO_2
    (4) NO_2

13. Which element consists of positive ions immersed in a "sea" of mobile electrons?

    (1) sulfur
    (2) nitrogen
    (3) calcium
    (4) chlorine

14. Which electron-dot diagram best represents a compound that contains both ionic and covalent bonds?

    (1) H:S̈:
         H

    (2) Ca²⁺ [:Ö: :Ö:S:Ö: :Ö:]²⁻

    (3) K⁺ [:B̈r:]⁻

    (4) :B̈r:B̈r:

15. Which is the correct electron-dot formula for a molecule of chlorine?

(1) $\text{Cl} \cdot \text{Cl}$ (2) $\cdot \overset{\cdot}{\underset{\cdot}{\text{Cl}}} \text{:} \overset{\cdot}{\underset{\cdot}{\text{Cl}}} \cdot$ (3) $\text{:} \overset{\cdot}{\underset{\cdot\cdot}{\text{Cl}}} \cdot \overset{\cdot}{\underset{\cdot\cdot}{\text{Cl}}} \text{:}$ (4) $\text{:} \overset{\cdot\cdot}{\underset{\cdot\cdot}{\text{Cl}}} \text{:} \overset{\cdot\cdot}{\underset{\cdot\cdot}{\text{Cl}}} \text{:}$

16. The electrical conductivity of KI(aq) is greater than the electrical conductivity of $H_2O(\ell)$ because the KI (aq) contains mobile

(1) molecules of H_2O (2) ions from H_2O
(3) molecules of KI   (4) ions from KI

17. Which compound contains both covalent bonds and ionic bonds?

(1) NaCl(s)   (2) HCl(g)   (3) $NaNO_3$(s)   (4) N_2O_5(g)

18. Which of the following compounds has the *least* ionic character?

(1) KI   (2) NO   (3) HCl   (4) MgS

19. A molecule of ammonia (NH_3) contains

(1) ionic bonds, only   (2) covalent bonds, only
(3) both covalent and ionic bonds   (4) neither covalent nor ionic bonds

20. Which bond has the greatest degree of ionic character?

(1) Li—Br   (2) F—F   (3) H—Cl   (4) S—O

21. Which electron dot formula represents a nonpolar molecule?

(1) $\text{H:}\overset{\text{H}}{\underset{\text{H}}{\text{C}}}\text{:}\overset{\cdot\cdot}{\underset{\cdot\cdot}{\text{Cl}}}\text{:}$ (2) $\text{H:}\overset{\text{H}}{\underset{\text{H}}{\overset{\cdot\cdot}{\text{N}}}}\text{:}$ (3) $\text{H:}\overset{\text{H}}{\underset{\text{H}}{\text{C}}}\text{:H}$ (4) $\text{H:}\overset{\cdot\cdot}{\underset{\text{H}}{\text{O}}}\text{:}$

22. Which formula represents a polar molecule?

(1) CH_4 (2) Cl_2 (3) NH_3 (4) N_2

MOLECULAR ATTRACTION

In addition to the forces responsible for the covalent bonds that hold the atoms together within individual molecules (intramolecular bonds), another set of forces holds the molecules to similar molecules (intermolecular bonds).

Molecules with No Dipoles (nonpolar molecules)

Molecules made up of only one kind of atom (H_2, N_2, P_4, S_8) are nonpolar. Electrons are shared equally among all the atoms.

Sometimes the shape of a molecule is such that a uniform distribution of charge results. Such a molecule is nonpolar, even though its individual

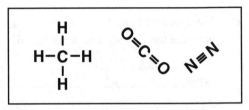

bonds are polar. For example, CO_2 is a nonpolar molecule held together by polar bonds. The arrangement of the atoms in this molecule can be represented O = C = O. The symmetrical shape of the molecule results in an even distribution of charge, making it nonpolar. Another example of a symmetrical molecule is CH_4 (tetrahedral).

Molecules with Dipoles (polar molecules)

The asymmetric, or unequal, distribution of electrical charge in most molecules causes one end of the molecule to be positive and the other end to be negative. This kind of molecule is called a **polar molecule** because it has a **dipole** (2 poles).

The force of attraction between the positive end of one dipole and the negative end of

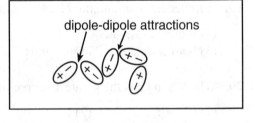

dipole-dipole attractions

a neighboring dipole helps to hold these molecules together within substances.

Some common examples of asymmetric molecules are H_2S (bent), PH_3 (pyramidal) and HCl (linear).

Hydrogen Bonding

When hydrogen atoms are covalently bonded to atoms of high electronegativity and small size (oxygen, nitrogen or fluorine), they behave almost as though they were hydrogen ions (bare protons). The reason for this is that they have a very small share of the bonding electron pair. This causes the hydrogen ends of the molecules to be especially attracted to the more electronegative atoms of neighboring molecules. These "super"

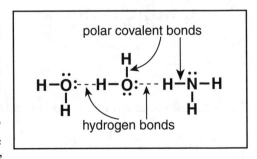

polar covalent bonds

hydrogen bonds

dipole-dipole attractive forces are known as **hydrogen bonds**. Hydrogen bonds account for some of the special properties of water, such as its unusually high boiling point. Other important molecules with hydrogen bonding are ammonia (NH_3), and hydrogen fluoride (HF). Hydrogen bonding is an example of a strong intermolecular force.

Weak Intermolecular Forces

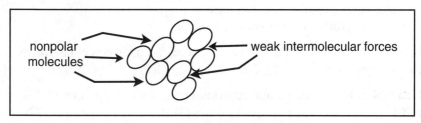

It has been found that even in nonpolar molecules, where there are no dipole attractions or hydrogen bonds, weak attractive forces exist among molecules. These forces make it possible for small, nonpolar molecules, such as hydrogen, helium, and nitrogen, to exist in the liquid (and even solid) phase under conditions of low temperature and high pressure. This is partially because these forces become more effective as molecules get closer to each other and they slow down. Weak intermolecular forces also increase as the number of electrons increases. Therefore, there is an increase in attraction with increasing molecular size.

Molecule-Ion Attraction

Another form of molecular attraction is the attraction between polar molecules and ions. It is this force that liberates the ions from their crystal lattice in ionic compounds as the compounds are dissolved in polar solvents. For example, when an ionic crystal, such as the salt, sodium chloride (NaCl), is added to water, a polar solvent, the sodium ions (Na^+) are attracted to the negative ends of water molecules. Chloride ions (Cl^-) are attracted to the positive ends of the molecules. The ions are freed from their crystal lattice and are then surrounded by water molecules, forming *hydrated ions*. The orienting of water molecules around ions is called *hydration* of the ions.

QUESTIONS

1. Which molecule is a dipole?   (1) H_2 (2) N_2 (3) CH_4 (4) HCl

2. Argon has a higher boiling point than neon because argon has
   (1) fewer electrons in its 2nd principal energy level
   (2) more electrons in its outermost principal energy level
   (3) weaker intermolecular forces of attraction
   (4) stronger intermolecular forces of attraction

3. Which of the following liquids has the weakest intermolecular forces of attraction between its molecules?   (1) Xe(ℓ)   (2) Kr(ℓ)   (3) Ne(ℓ)   (4) He(ℓ)

4. Which molecule is nonpolar and contains a nonpolar covalent bond?
   (1) CCl_4 (2) F_2 (3) HF   (4) HCl

5. The strongest hydrogen bonds are formed between molecules in which hydrogen is covalently bonded to an element with
   (1) high electronegativity and large atomic radius
   (2) high electronegativity and small atomic radius
   (3) low electronegativity and large atomic radius
   (4) low electronegativity and small atomic radius

6. Weak intermolecular forces of attraction between molecules always decrease with
   (1) increasing molecular size and increasing distance between the molecules
   (2) increasing molecular size and decreasing distance between the molecules
   (3) decreasing molecular size and increasing distance between the molecules
   (4) decreasing molecular size and decreasing distance between the molecules

7. In which liquid is hydrogen bonding the most significant force of attraction?
   (1) HF(ℓ)   (2) HCl(ℓ)   (3) HBr(ℓ)   (4) HI(ℓ)

8. The unusually high boiling point of H_2O is primarily due the presence of
   (1) hydrogen bonds   (2) ionic bonds
   (3) weak intermolecular forces   (4) molecule-ion attractions

9. At 298 K, the vapor pressure of H_2O is less than the vapor pressure of CS_2. The best explanation for this is that H_2O has
   (1) larger molecules   (2) a larger molecular mass
   (3) stronger ionic bonds   (4) stronger intermolecular forces

10. Which formula represents a tetrahedral molecule?
    (1) CH_4 (2) $CaCl_2$ (3) HBr   (4) Br_2

11. Molecule-ion attractions are present in
    (1) NaCl(aq)   (2) HCl(g)   (3) CCl_4(ℓ)   (4) $KClO_3$(s)

12. As the distance between molecules of a liquid increases, the weak intermolecular forces of attraction   (1) decrease   (2) increase   (3) remain the same

PRACTICE FOR CONSTRUCTED RESPONSE

1. **a.** Explain why water has such a high melting point. (1point)

_____

_____

b. Draw a sketch of water molecules as a solid showing their interactive forces. (1 point)

2. In the bonding of KCl

a. What kind of bond forms between K and Cl? (1 point)

_____

b. Give the electron dot symbols for K and Cl as elements. (1 point)

_____

c. Draw the electron dot structure for the compound KCl. (1 point)

3. In the bonding of NH_3

a. What kind of bonds form between N and H? (1 point)

_____

b. Give the electron dot symbols for N and H as elements. (1 point)

_____

c. Draw the electron dot structure for the compound NH_3. (1 point)

4. Explain, using electron dot notation, why a Noble Gas such as neon tends not to bond. (3 points)

_____

_____

_____

5. For metallic bonding

    a. Name an element that has this kind of bonding. (1 point)

_____

    b. Sketch a model of atoms metallicly bonded. (1 point)

    c. Use your sketch to explain why metallicly bonded elements have *high density* and *good electrical conductivity*. (2 points)

_____

_____

6.

$$N \equiv N$$

$$O \overset{\displaystyle H}{\underset{\displaystyle H}{\diagup}}$$

$$O = C = O$$

For each of the three molecules above, indicate if it is polar or non-polar, and give the reason for your answer. (6 points)

a. _____

_____

b. _____

_____

c. _____

_____

CHAPTER 5
PHYSICAL BEHAVIOR OF MATTER

In studying chemistry we study the composition of matter, and how it changes. Energy changes accompany these changes in matter.

ENERGY

Although there are many more sophisticated definitions, **energy** is probably best defined as the capacity to do work. **Work** is done whenever a force is used to change the position and/or motion of matter in a system.

All energy can be classified into two types—*potential energy* and *kinetic energy*. **Potential energy** is stored energy. **Kinetic energy** is energy of motion.

When we roll a ball up a hill, we expend kinetic energy. The ball, however, acquires potential energy as a result of the work done on it. At the top of the hill, the ball at rest has potential energy due to its position. If the ball rolls down the hill, the potential energy will change to the kinetic energy of motion.

Forms of Energy

Energy, whether potential or kinetic, is associated with physical phenomena. Traditionally, different *forms* of energy are identified as follows:

- **Thermal energy** is the total amount of internal energy an object has due to the random motion of its molecules.
- **Chemical energy** is the energy associated with chemical change involving the making and breaking of bonds.
- **Light energy** is the energy associated with electromagnetic radiation (visible, ultraviolet, infrared, x-rays, etc.)
- **Nuclear energy** is the energy associated with changes in the masses of atoms' nuclei.
- **Heat energy** is a transfer of energy (usually thermal) from a body of higher temperature to a body of lower temperature.
- **Electrical energy** and **mechanical energy** are forms less important in chemical reactions.

Conservation of Energy

During ordinary chemical or physical changes, energy is conserved. Energy may be changed from one form to another, or transferred from one body or system to another, but the total amount of energy remains constant.

Any activity that releases energy is said to be **exothermic.** Activities that require energy are said to be **endothermic**.

Measurement of Energy

Temperature is a measure of the *average* kinetic energy of the particles in a substance. Temperature is not a form of energy. Two things should be emphasized in this definition:

1. The word *average* indicates that temperature might be calculated by adding the kinetic energies of all the particles and dividing by the number of particles. Under most conditions, each particle possesses its own kinetic energy, which may be quite different from that of any of its neighboring particles.

2. The word *heat* does *not* appear in the definition, although the temperature may be affected by the amount of heat in the system. A rising temperature is associated with the addition of heat to the system; falling temperature is associated with a loss of heat. Whenever two bodies with different temperatures get close enough to each other, *heat* flows from the body with the higher temperature to the one with the lower temperature until the two bodies are at the same temperature.

A **calorimeter** is a device used to measure heat of reaction. The calorimeter uses the change in temperature of water to measure the amount of heat produced during exothermic processes and the amount of heat absorbed during endothermic processes. The process is allowed to proceed in a container that is in contact with a known mass of water. The temperature of the water is measured before and after the reaction (process), and the change in temperature is determined.

Temperature change must be converted to units of heat energy. A traditional unit of heat energy is the **calorie**. One calorie is defined as the quantity of heat needed to raise the temperature of 1 gram of water by 1 Celsius degree. Using standard units, the amount of heat required to raise 1 gram of water by 1 Celsius degree is 4.18 **Joules**.

A conversion constant known as the *specific heat* can make the same equation work for either unit and for substances other than water. For water, the specific heat is either 1.00 calorie/gram°C or 4.18 joule/g.°C.

The equation is: $q = m \, C \, \Delta T$ where: q = heat gained or lost
m = mass
C = specific heat
ΔT = temperature change

SAMPLE PROBLEMS

1. How many Joules are absorbed by 30.0g of water when its temperature is raised from 20.°C to 40.°C?

Solution:

Given:	mass (water) = 30.0g
	$\Delta T = 40.°C - 20.°C = 20.°C$
Known:	C = 4.18 J/g. °C (for water)
Find:	heat gained
Equation:	Heat gained = mCΔT
	$= 30.0g \times 4.18 \text{ J/g°C} \times 20.°C$

$$= 2.5 \times 10^3 \text{ J}$$

2. A reaction chamber inside a calorimeter contains 150.g of water at 19°C. A reaction is permitted to take place in the chamber. After the reaction, the temperature of the water is 29°C. How many calories were released by the reaction?

Heat released = mass × specific heat × temperature change

= mCΔT

$= 150.g \times 1 \text{ cal/g°C} \times 10.°C$

$$= 1.5 \times 10^3 \text{ calories}$$

QUESTIONS

Answer the following questions using Tables B, C, and D of the *Reference Tables for Physical Setting/Chemistry.*

1. The temperature of a substance is a measure of its particles'

(1) average potential energy   (2) average kinetic energy
(3) enthalpy   (4) entropy

2. How many kiloJoules (kJ) are equivalent to 1.0 Joule?

(1) 0.001 kJ   (2) 0.01 kJ   (3) 1000 kJ   (4) 10,000 kJ

3. What is the total number of Joules of heat energy absorbed when the temperature of 200. grams of water is raised from 10°C to 40°C?

    (1) 125 J   (2) 836 J   (3) 25000 J   (4) 33400 J

4. How many grams of water will absorb a total of 5.02×10^3 Joules of energy when the temperature of the water changes from 10.0°C to 30.0°C?

    (1) 10.0 g   (2) 20.0 g   (3) 30.0 g   (4) 60.0 g

5. How many Joules of heat energy are absorbed in raising the temperature of 10. grams of water from 5.0°C to 20.°C?

    (1) 6.3×10^2 (2) 2.5×10^2 (3) 1.2×10^2 (4) 3.0×10^1

6. The temperature of 50 grams of water was raised to 50°C by the addition of 4200 Joules of heat energy. What was the initial temperature of the water?

    (1) 10°C   (2) 20°C   (3) 30°C   (4) 60°C

MATTER

The traditional definition of **matter** is anything that has mass and volume. The term matter is ordinarily used when referring to any material found in nature. These materials are found in what seems to be an infinite variety of forms, and are classified in many ways. Two groups into which all matter may be placed are *homogeneous matter* and *heterogeneous matter.*

- **Homogeneous matter** includes all those materials having uniform characteristics throughout a given sample. This group includes elements, compounds, and solutions.
- **Heterogeneous matter** is not uniform in its composition. Its parts are not alike.

Scientists also classify matter into the two groups—*pure substances* and *mixtures* of substances. **Pure Substances** are almost always homogeneous. **Mixtures** may be homogeneous or heterogeneous. Solutions are examples of homogeneous mixtures.

Pure Substances

Pure substances can be elements or compounds. Elements are composed of atoms, all of which have the same atomic number. They cannot be broken down into other substances by chemical means. Examples are O_2 and Ne. Compounds are made of two or more elements but have a definite composition of elements. Examples of compounds are H_2O and CO_2. Pure substances have constant composition and constant properties which can be used in their identification. (Examples of such properties are: melting point, boiling point, solubility, and chemical reactivity.)

PHASES OF MATTER

Matter exists in nature in three **phases**—solid, liquid, and gas. These terms might be used to describe materials as they are found in nature under ordinary conditions. Water is most often found in liquid form, air is a mixture of gases, and the rocky portion of the earth consists mostly of materials in solid form.

In the consideration of the phases of matter, we shall be concerned with the transfer of heat as different materials undergo *phase changes*—that is, changes from one phase to another. A change in temperature or pressure can make a pure substance change from one phase to another without changing its chemical identity. The kinetic molecular theory of gases can be used not only to model gases, but it can also explain the properties of solids and liquids.

Kinetic Molecular Theory of Gases. Since so many gases exhibit the same behavior, a model was developed to help explain this similarity. This kinetic molecular theory of gases is based on the following assumptions:

1. All gases are composed of tiny, individual particles called molecules that are in continuous motion. These particles move rapidly, randomly and in straight lines.
2. When particles collide with one another, energy is transferred *without loss,* from one particle to the other. Therefore, the net total energy of the system remains constant.
3. Compared to the distances between them, the particles are so small that their volumes are considered to be zero.
4. The particles have no attraction for one another.

Gases that have these properties are known as **ideal gases.**

PROPERTIES OF GASES, LIQUIDS AND SOLIDS

Gases

Gases are transparent, they can be compressed and they expand without limit. Gases assume the shape and volume of their container. The volume of a given mass of a gas is the volume of the container that holds it.

Gases exert pressure on the walls of a container. Pressure is most often expressed in units of force per unit area of surface.

Earth's atmosphere exerts pressure, called air pressure, or atmospheric pressure. Air pressure is measured with a barometer. One type of barometer consists of a glass tube about one meter long, sealed at one end and partly filled with mercury. The open end of the tube is placed in a reservoir of mercury that is exposed to the atmosphere. The weight of the air exerts a pressure on the surface of the mercury in the reservoir. This pressure supports a column of mercury in the glass tube.

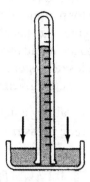

Figure 5-1. Mercury Barometer. The greater the air pressure (arrows) the higher the mercury rises in the tube.

The height of the mercury column is directly related to the amount of pressure exerted by the atmosphere on the mercury in the reservoir. At 0°C at sea level, the atmospheric pressure supports a column of mercury 760 mm high. In honor of E. Torricelli, who invented the mercury barometer, this pressure is often called 760 torr. This value—760 torr or 760 mm of mercury—is also known as **one atmosphere (1.00 atm)** of pressure, or **standard pressure (STP)**. It is equal to **101.3 kilopascals (kPa)**.

The kinetic molecular theory of gases explains pressure as the force of collision of the moving gas molecules against the walls of the container. Gases are greatly compressible because of the empty spaces which exist between the molecules. Gases assume their container's shape because the molecules of gases are not connected.

Liquids

Liquids have definite volumes but do not have definite shapes. Like gases, liquids take on the shape of their containers. Liquids, however, have low compressibility.

The kinetic molecular theory explains these properties. Liquids have low compressibility because the distances between molecules are not large. The molecules of liquids are packed close together, virtually touching each other. Liquids do not have their own shapes because the molecules are not rigidly connected to each other.

Evaporation. The force of attraction among particles of a liquid is much greater than among particles of a gas. The particles in a liquid are in constant motion, their speed depending on temperature. Particles that have a greater-than-average speed can overcome forces of attraction and leave the surface of the liquid to enter the gas, or vapor, phase. The term **vapor** is used to refer to the gas phase of a substance that is usually a solid or liquid at room temperature.

The escape of particles from the surface of a liquid to form the vapor is called **evaporation.** In an open container, evaporation continues until all the liquid has vaporized.

Condensation. If a liquid is placed in a closed container at a given temperature, only a small quantity will evaporate. It is impossible for particles that escape into the gas phase to leave the container. As more and more particles enter the gas phase, the likelihood that some of them will strike the liquid surface and return to the liquid phase increases. The change from vapor to liquid is called **condensation.** In a closed system, the rate of condensation eventually becomes equal to the rate of evaporation at the given temperature.

Vapor Pressure. In any closed system containing a liquid, the vapor produced by evaporation exerts a pressure, which is called the **vapor pressure** of the liquid. Every liquid has its own characteristic vapor pressure. Vapor pressure increases as temperature increases.

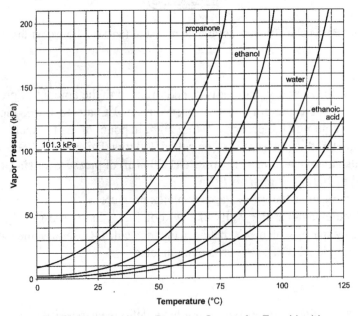

Figure 5-2. Vapor Pressure Curves for Four Liquids.

Boiling Point. Upon being heated, most liquids eventually boil. Boiling is characterized by the formation of bubbles of vapor in the body of the liquid, as the change from liquid to vapor takes place within the body of the liquid, as well as at the surface.

As described earlier, vapor pressure increases with an increase in temperature. The temperature at which the vapor pressure of a liquid is equal to the pressure pushing down on the surface of the liquid is called the **boiling point** of the liquid. The **normal boiling point** of a liquid is defined as the temperature at which the vapor pressure of the liquid is one atmosphere, or 760 torr or 101.3 kPa.

In most circumstances, when reference is made to the boiling point of a substance, it is the normal boiling point that is indicated. However, the temperature at which a liquid will boil can be changed by changing the pressure pushing down on the surface of the liquid. For example, the vapor pressure of water at 80°C is 47 kPa. Thus, if the pressure on the surface of a sample of water is reduced to 47 kPa, the water will boil at 80°C. Conversely, increasing the pressure raises the boiling point temperature.

Heat of Vaporization and Heat of Condensation. The particles that escape from the surface of a liquid are those having the highest kinetic energy. The remaining particles have lower average kinetic energy. Thus, heat must be added to the liquid to maintain constant temperature. Vaporization, then, is an endothermic process.

$$heat + H_2O(\ell) \rightarrow H_2O(g)$$

The **heat of vaporization** of a liquid is the number of calories per gram of liquid that must be added in order to maintain a constant temperature while vaporization, or boiling, occurs. For water at its normal boiling point of 100°C, the heat of vaporization is 539 calories per gram, or 2259 Joules per gram.

Heat of condensation is the number of calories per gram that must be *removed* in order to maintain constant temperature during condensation. Steam condensing on your hand produces severe burns due to the heat liberated on condensation. Steam at 100°C is much more damaging than water at the same temperature. The process of condensation is exothermic.

$$H_2O(g) \rightarrow H_2O(\ell) + heat$$

For water, the heat of condensation is 539 calories per gram or 2259 Joules per gram – the same as the heat of vaporization.

Solids

Solids have a definite volume and a definite shape. The particles in a solid have an orderly arrangement in regular geometric patterns, called crystals. Some apparent solids (referred to as amorphous solids - super cooled liquids) do not have regular geometric patterns. Some examples are: glass, plastics, and wax. Although the particles in a solid do not move freely about throughout the solid, they are in constant motion. They vibrate around their fixed positions. The molecules of solids are close together, similar to liquids, so they are only slightly more compressible than liquids.

Freezing-Melting Point. When a liquid is cooled (heat is removed), a temperature is eventually reached at which the liquid begins to freeze. It changes to a solid. This temperature, which remains constant until all the liquid has solidified at 1 atmosphere pressure, is called the **freezing point** of the liquid. While the liquid is cooling, the average kinetic energy of its particles decreases until it is low enough for the attractive forces to be able to hold the particles in the fixed positions characteristic of the solid phase.

Alternately, warming a solid eventually causes the solid to melt. It begins to change to a liquid. During this phase change, the temperature remains constant until all the solid has liquefied at 1 atmosphere pressure. This temperature is called the **melting point** of the solid. For any given substance, the melting point temperature is exactly the same as the freezing point temperature. The only difference is in the direction of approach.

The melting-freezing point of a substance may also be defined as the temperature at which the liquid phase and the solid phase exist in equilibrium.

Heat of Fusion and Heat of Solidification. Melting is an endothermic process. In order to maintain a constant temperature during this phase change, heat must be continually added to the system.

$$\text{heat} + H_2O(s) \rightarrow H_2O(\ell)$$

The amount of heat needed to change a unit mass of a substance from solid to liquid at constant temperature and 1 atmosphere pressure is called the **heat of fusion** of that substance. The heat of fusion of ice at 0°C and 1 atmosphere is 334 Joules per gram. To change one gram of ice at 0°C and 1 atmosphere pressure to one gram of water at the same conditions, 334 Joules of heat must be added. Conversely, to change 1 gram of liquid water at 0°C and 1 atm to 1 gram of ice at the same conditions, 334 Joules of heat must be removed. This is the **heat of solidification.**

$$H_2O(\ell) \rightarrow H_2O(s) + \text{heat}$$

Because of the principle of the conservation of energy, the heat of solidification of a substance is equal to the heat of fusion. If 334 J/g. must be added to melt one gram of ice, we must remove 334 J/g to refreeze it.

A graph of temperature vs time will make it easier to visualize the phase change process. Figure 5-3 is such a graph for a substance with a melting point of 50°C and a boiling point of 110°C. The portion of the curve between A and B represents the time the solid phase is being heated. Note the uniform increase in temperature during this interval. The portion of the curve between B and C represents the time the substance is undergoing a phase change, from solid to liquid (heat of fusion). Note that there is no change in temperature during this interval, even though heat is being added at a constant rate. The heat is being absorbed in the process of the phase change.

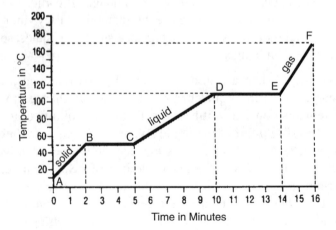

Figure 5-3. Heating Curve

The part of the curve between C and D represents the time interval during which the liquid is being heated. Note the uniform increase in temperature during this interval. At point D, the temperature stops rising.

The interval between D and E represents the time interval in which the liquid is changing to a gas (heat of vaporization). Once again, there is no change in temperature during this time. At point E, all of the liquid has changed to a gas, and the temperature of the gas begins to rise at a uniform rate.

Such a graph is called a *heating curve.* If you were to read this graph from right to left, it would be a *cooling curve.*

Sublimation. Under certain conditions, it is possible for a substance to change from a solid directly into the gas phase without obviously passing through the liquid phase. This solid-to-gas change is called **sublimation.** Iodine and carbon dioxide are examples of substances that may undergo sublimation. Iodine and carbon dioxide have small non-polar molecules with weak intermolecular attractions.

QUESTIONS

Answer the following questions using Tables A, B, and H of the *Reference Tables for Physical Setting/Chemistry.*

1. The heat of vaporization for water at its normal boiling point is

(1) 0.069 kJ/g (2) 0.25 kJ/g (3) 10.6 kJ/g (4) 2.26 kJ/g

2. Which conditions of pressure and temperature exist when ice melts at its normal melting point?

(1) 1 atm and 0°C   (2) 760 atm and 0°C
(3) 1 atm and 0 K   (4) 760 atm and 273 K

3. The graph below represents the uniform cooling of water at 1 atmosphere, starting with water as a gas above its boiling point.

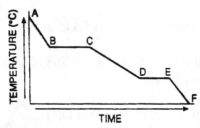

Which segments of the cooling curve represent the freezing temperature?

(1) AB and CD   (2) DE   (3) AB and EF   (4) CD and EF

4. The heat of fusion of a substance is the energy measured during a

(1) phase change   (2) temperature change
(3) chemical change   (4) pressure change

5. Which material exists as a supercooled liquid at STP?

(1) salt   (2) sand   (3) diamond   (4) glass

6. Which substance will sublime at room temperature (20°C)?

(1) $C_{12}H_{22}O_{11}(s)$ (2) $C_6H_{12}O_6(s)$
(3) $SiO_2(s)$ (4) $CO_2(s)$

7. When the temperature of a sample of water is changed from 20°C to 100°C, the change in its vapor pressure is

(1) 2.7 kPa   (2) 93 kPa
(3) 98 kPa   (4) 101.3 kPa

8. As ice at 0°C changes to water at 0°C, the average kinetic energy of the ice molecules

(1) decreases   (2) increases   (3) remains the same

9. Which process occurs when dry ice, $CO_2(s)$, is changed into $CO_2(g)$?

(1) crystallization   (2) condensation
(3) sublimation   (4) solidification

10. As water in a sealed container is cooled from 20°C to 10°C, its vapor pressure

(1) decreases   (2) increases   (3) remains the same

11. Water will boil at a temperature of 40°C when the pressure on its surface is

(1) 1.9 kPa   (2) 2.2 kPa   (3) 7.3 kPa   (4) 101.3 kPa

12. A 1-gram sample of which substance in a sealed 1-liter container will occupy the container completely and uniformly?

(1) Ag(s)   (2) $Hg(\ell)$ (3) $H_2O(\ell)$ (4) $H_2O(g)$

Base your answer to questions 13 and 14 on the graph below, which represents the uniform heating of a water sample at standard pressure, starting at a temperature below 0°C.

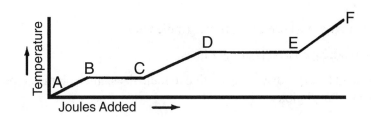

13. The number of Joules required to vaporize the entire sample of water at its boiling point is represented by the interval between

    (1) A and B   (2) E and F   (3) C and D   (4) D and E

14. If 5.0 grams of water undergoes a temperature change from C to D, the total energy absorbed is

    (1) 340 Joules   (2) 420 Joules   (3) 760 Joules   (4) 2100 Joules

15. The vapor pressure of ethanol at its normal boiling point is

    (1) 10.7 kPa   (2) 13.3 kPa   (3) 36.4 kPa   (4) 101.3 kPa

16. Which substance *cannot* be decomposed by a chemical change?

    (1) mercury (II) oxide   (2) potassium chlorate
    (3) water   (4) copper

17. Which change of phase represents fusion?

    (1) gas to liquid   (2) gas to solid
    (3) solid to liquid   (4) liquid to gas

18. Which is the equivalent of 750. Joules?

    (1) 0.750 kJ   (2) 7.50 kJ   (3) 75.0 kJ   (4) 750. kJ

IDEAL GAS LAWS

Ideal gases are those whose gas particles (molecules):

1. travel in random, constant, straight lines.

2. are separated by great distances relative to their sizes; the volume of the gas particles is considered negligible.

3. have no attractive forces between them.

4. have collisions that may result in a transfer of energy between particles, but the total energy of the system remains constant.

The properties of gases can be readily explained with the kinetic molecular theory. The thermal energy of gas molecules is due to their random motion. The temperature is their average kinetic energy which depends on both their mass and speed. The more mass or speed, the greater the average kinetic energy of the molecules (or temperature).

If a constant amount of a gas like oxygen is heated, its mass cannot change so the molecules of oxygen must go faster to be at a higher temperature. If they go faster, they should collide harder and more frequently on the sides of the container. Therefore the higher the **temperature**, the higher the **pressure** for a constant volume.

If we want the pressure to stay constant when the temperature increases, the volume must become larger so that faster moving molecules are farther apart. Even though the molecules at higher temperature are moving faster, their lower concentration compensates so the pressure can stay constant. As a result, as the **temperature** of a gas increases, for it to remain at constant pressure, its **volume** must increase.

At constant temperature if the volume of a gas is decreased, the concentration of molecules increases making more frequent collisions. As a result, as the **volume** decreases the **pressure** increases.

At 0°C (the freezing point of water) the molecules of water are still moving. They can go slower, so 0°C cannot be **absolute zero**, where there is no kinetic energy of the molecules. Absolute zero is 0 Kelvin, or 0 K. To change to the Kelvin (absolute) temperature scale 273° must be added to the Celsius temperature

$$K = °C + 273°$$

The freezing point of water is 273 K and the boiling point is 373 K at one atmosphere. In all gas law problems the temperature must be in Kelvin degrees.

The properties can be combined into one equation known as the **combined gas law** for an ideal gas.

$$\frac{P_1 V_1}{T_1} = \frac{P_2 V_2}{T_2}$$

Note:

If temperature is constant $T_1 = T_2$, the equation becomes $P_1 V_1 = P_2 V_2$

If the pressure is constant $P_1 = P_2$, the equation becomes

$$\frac{V_1}{T_1} = \frac{V_2}{T_2}$$

If the volume is constant $V_1 = V_2$ and the equation becomes

$$\frac{P_1}{T_1} = \frac{P_2}{T_2}$$

The advantages of using this equation are:

- Only one equation is needed

- All initial values are on the left of the equation; all new values are on the right of the equation.

- In cases where either temperature or pressure is unchanged, it can be ignored (or included); its value will not affect the problem.

It is also recommended that tables be set up and used. The following model can be used:

	1 Initial conditions	2 New or final conditions
Volume		
Pressure		
Temperature K		

SAMPLE PROBLEM

A sample of gas had a volume of 200. L under a pressure of 0.300 atm and a temperature of 20.° C. What volume would the sample occupy at a pressure of 0.250 atm and a temperature of 30.°C?

Solution:

1. Convert temperatures to Kelvin:

   $20.°C + 273 = 293K$; $30.°C + 273 = 303K$

2. Set up a table.

	1	2
V	200. L	x
P	0.300 atm	0.250 atm
T	293 K	303K

3. Write equation:

   $$\frac{V_1 P_1}{T_1} = \frac{V_2 P_2}{T_2} \text{ or } V_2 = V_1 \times \frac{P_1}{P_2} \times \frac{T_2}{T_1}$$

4. Substitute:

   $$V_2 = 200. \text{ L} \times \frac{0.300 \text{ atm}}{0.250 \text{ atm}} \times \frac{303K}{293K}$$

   $$= 248 \text{ L}$$

Avogadro's Hypothesis. Two equal volumes of gases at the same temperature and pressure must have equal numbers of particles, even though their masses are not the same. For example, at the same temperature and pressure, one liter of nitrogen gas contains the same number of molecules as one liter of hydrogen gas, even though the mass of the nitrogen gas is fourteen times that of the hydrogen gas.

Since the two gases are at the same temperatures, they cannot have the same average speeds. Since the heavier nitrogen molecules exert the same pressure as the lighter hydrogen, the nitrogen must be moving at a slower speed. At the same temperature, lighter molecules travel faster than heavier molecules.

Deviations from the Gas Laws. Real gases do not behave exactly as predicted by the "ideal" gas of the model. Deviations from the ideal behavior are due to the fact that gas particles *do have volume* and they *do exert some attraction* for one another. Deviations from the gas laws are least obvious among light gases at high temperatures and low pressures. These conditions are optimum for high kinetic energies and maximum separation of particles. Hydrogen and helium are closest to "ideal" gases.

Standard Temperature and Pressure. Gas volumes are influenced considerably by changes in temperature and pressure. Thus, when working with gases, it is convenient to define standard reference conditions. By convention, 0°C (273K) and 760 torr or 101.3 kPa (1 atm) represent **standard temperature** and **pressure**, often designated as **STP**. When dealing with a gas volume, STP is assumed unless otherwise indicated.

$$\boxed{\textbf{QUESTIONS}}$$

Answer the following questions using Tables A, B, C, and D of the *Reference Tables for Physical Setting/Chemistry*.

1. The boiling point of water at standard pressure is

    (1) 0.000 K   (2) 100. K   (3) 273 K   (4) 373 K

2. Which temperature represents absolute zero?

    (1) 0 K   (2) 0°C   (3) 273 K   (4) 273°C

3. A 100.-milliliter sample of helium gas is placed in a sealed container of fixed volume. As the temperature of the confined gas increases from 10.°C to 30.°C, the internal pressure

    (1) decreases   (2) increases   (3) remains the same

4. If the pressure on 36.0 milliliters of a gas at STP is changed to 0.190 atm at constant temperature, the new volume of the gas is

    (1) 9.00 mL   (2) 126 mL   (3) 189 mL   (4) 226 mL

5. At constant temperature the pressure on 8.0 liters of a gas is increased from 1 atmosphere to 4 atmospheres. What will be the new volume of the gas?

(1) 1.0 L   (2) 2.0 L   (3) 32 L   (4) 4.0 L

6. Which sample of methane gas contains the greatest number of molecules at standard temperature?

(1) 22.4 liters at 1 atmosphere   (2) 22.4 liters at 2 atmospheres
(3) 11.2 liters at 1 atmosphere   (4) 11.2 liters at 2 atmospheres

7. The pressure on 200. milliliters of a gas is decreased at constant temperature from 0.900 atm to 0.800 atm. The new volume of the gas, in milliliters, is equal to

(1) $200. \times \dfrac{0.900}{0.800}$ (2) $200. \times \dfrac{0.800}{0.900}$

(3) $0.800 \times \dfrac{200.}{0.900}$ (4) $0.800 \times \dfrac{0.900}{200.}$

8. The diagram below represents a gas confined in a cylinder fitted with a movable piston.

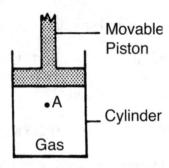

As the piston moves toward point A at a constant temperature, which relationship involving pressure (P) and volume (V) is correct?

(1) $P + V = k$ (2) $P - V = k$ (3) $P \div V = k$ (4) $P \times V = k$

9. What would be the volume at STP of 255 mL of CO_2 measured at 30.°C and 85.0 kPa?

(1) 193 mL   (2) 1950 mL   (3) 337 mL   (4) 237 mL

10. Given: 400 mL of gas at STP. What is the pressure at 320 K if the volume is 500. mL?

    (1) 0.683 atm   (2) 0.938 atm   (3) 1.07 atm   (4) 1.47 atm

11. A gas occupies a volume of 560 mL at a temperature of 100°C. To what temperature must the gas be changed if it is to occupy 400 mL, with the pressure remaining unchanged?

    (1) 71.4 K   (2) 522 K   (3) 100 K   (4) 266 K

12. The pressure of a gas in a steel can is 92.0 kPa at 25°C. What will be the pressure if the temperature is increased to 50.°C?

    (1) 46 kPa   (2) 184 kPa   (3) 99.7 kPa   (4) 84.9 kPa

MIXTURES

Matter can exist as pure substances or mixtures. Mixtures are composed of two or more different substances that can be separated by physical means. The proportions of components in a mixture can be varied. Each component retains its original properties.

The physical separation of the components of the mixture depends on the differences in their physical properties such as density, practicle size, molecular polarity, boiling point, freezing point, and solubility. Some of the processes used to separate mixtures are described below.

Filtration. A mixture is poured through a porous material such as paper or fabric. It relies on the different sizes of the components in the mixture to make a separation. The smaller, usually molecular, component passes through the pores, but the larger components cannot. For example, if muddy water is poured into filter paper, the water molecules pass though, but the mud particles are caught in the paper.

Distillation. This method relies on the different boiling points of the mixture's components. The materials, which must be solids or liquids, are heated in a distilling flask. The lowest boiling point component boils first, leaving the other components in the flask. The vapor of the first component can be captured and condensed back into a liquid.

Chromotography. This method relies on the different abilities of components to diffuse through a material. Such abilities to diffuse may depend on the component's polarity or molecular size. In paper chromatography a solvent (such as water) can carry along components (such as water soluble ink pigments) as the solvent moves up through the paper by capillary action. Some pigments move farther than others.

Mixtures are either *heterogeneous* or *homogeneous*.

Heterogeneous mixtures are not uniform in the distribution of their components. Heterogeneous mixtures can usually be separated by filtration or by density differences. An example would be muddy water. If allowed to sit undisturbed, eventually the mud will settle to the bottom. The mud could also be filtered out.

Homogeneous mixtures have their components uniformly distributed so that the mixture is the same in all parts. The components cannot be separated as easily as those in heterogeneous mixtures. They cannot be filtered out and do not settle out. Distillation and chromatography are often successful techniques for separating homogeneous components. Solutions are examples of homogeneous mixtures.

SOLUTIONS

A solution is defined as a homogeneous mixture of two or more substances. The substance that is dissolved is called the **solute.** The substance in which the solute is dissolved is called the **solvent.**

In beginning chemistry courses, most solutions have water as the solvent. All such solutions are called **aqueous solutions**.

The most common solutions consist of solid solutes dissolved in liquid solvents. Solutions consisting of two liquids are also fairly common. In such solutions, the liquid present in excess is considered to be the solvent.

Some liquids cannot be mixed together to form solutions. Such liquids are said to be *immiscible.* A mixture of oil and water is an example of immiscible liquids. Liquids that will form homogeneous mixtures in almost any proportions, such as methyl alcohol and ethyl alcohol, are said to be *miscible.* Mixtures of miscible liquids are true solutions.

Methods of Indicating Solubility of Solutes

The **solubility** of a substance in a given solvent refers to how much of the substance can be dissolved in a given quantity of the solvent at a specified temperature and pressure. Words like *soluble* and *insoluble* are used to indicate that relatively very large or very small amounts of the solute will dissolve. Certain tables use specified terms to describe *ranges* of solubility. Terms like nearly insoluble, slightly soluble, soluble, and very soluble are often used.

Sometimes solubility curves are provided showing the maximum quantity of solute dissolved in l00g of solvent plotted against temperature. (Such as in Table G of the *Reference Tables for Physical Setting/Chemistry*.)

Saturated solutions are solutions which contain the maximum solute that will dissolve in a given quantity of solvent at a specified temperature.

Unsaturated solutions contain less than this amount of solvent.

Supersaturated solutions contain more than this amount of solute.

A supersaturated solution can be produced by dissolving at elevated temperatures and slowly allowing the solution to cool to room temperature. Such solutions tend to be very unstable. The excess solute will suddenly "fall out" of solution if the solution is agitated or a few seed particles of solute are added. The term *precipitation* is used to describe an event in which dissolved material falls out of solution. Usually this material settles to the bottom of the container.

Methods of Indicating Concentration

A variety of methods are used to designate the concentration of solutes in solutions. These methods use different units and different relationships between solute and solvent, or between solute and solution.

Some of the more familiar terms used to indicate concentration are mainly descriptive, and are of little use quantitatively. These terms include concentrated, dilute, weak and strong. **Dilute** solutions contain small amounts of solute; **concentrated** solutions contain large amounts of solute. **Weak** solutions are dilute; **strong** solutions are concentrated.

Molarity. The **molarity (M)** of a solution is one of several terms used to indicate the amount of solute in a given amount of solution. A one molar solution (1. 00**M**) contains one mole of *solute* dissolved in enough solvent to make 1.00 liter of *solution*. A general expression for molarity is:

$$\mathbf{M} = \frac{\text{number of moles of solute}}{\text{liters of solution}}$$

A two molar (2**M**) solution contains 2 moles of solute per liter of solution. A 0.1 molar (0.1**M**) solution contains 0.1 mole per liter of solution.

SAMPLE PROBLEM

What is the molarity of a solution of sodium oxalate, $Na_2C_2O_4$, containing 33.5 grams of solute in 100.0 mL of solution?

Solution:

1. Calculate the mass of solute in one liter of solution:

$$\frac{33.5}{100mL} \times 1000 \text{ mL / liter} = 335 \text{ g / liter}$$

2. Calculate the mass of one mole of $Na_2C_2O_4$:

$$\text{mole mass} = (2 \times 23) + (2 \times 12) + (4 \times 16) = 143 \text{ g}$$

3. Calculate the number of moles of solute per liter of solution (definition of molarity):

$$\frac{335 \text{ g}}{1 \text{ liter}} \times \frac{1 \text{ mole}}{143 \text{ g}} = 2.50 \text{ moles / liter} = \textbf{2.50 M}$$

If the molarity of a solution is known, the number of moles in a given volume of that solution can be calculated as follows:

$$\text{moles of solute} = \text{molarity} \times \text{volume in liters}$$

SAMPLE PROBLEM

How many moles of NaOH are contained in 200mL of a 0.1M solution of NaOH?

Solution:

1. Find the volume of the solution in liters:

$$200 \text{ mL} \times \frac{1 \text{ liter}}{1000 \text{ mL}} = 0.2 \text{ liter}$$

2. Substitute this volume into the equation:

$$\text{moles of solute} = 0.1 \text{ moles/liter} \times 0.2 \text{ liter}$$
$$= \textbf{.02 moles NaOH}$$

Percent by Mass. Percent by mass is calculated by dividing the mass of the solute by the mass of the solution (total mass of solvent + solute) and multiplying by 100%.

$$\% \text{ (mass)} = \frac{\text{mass of solute}}{\text{mass of solution}} \times 100\%$$

SAMPLE PROBLEM

What is the percent by mass concentration of a solution where 5.0 g of salt are dissolved in 80. g of water?

Solution:

1. Find the mass of the solution:

$$5.0 \text{ g} + 80. \text{ g} = 85 \text{ g}$$

2. Multiply the ratio of solute to solvent by 100%:

$$\frac{5.0 \text{ g}}{85 \text{ g}} \times 100\% = \textbf{5.9\%}$$

Percent by Volume. Percent by volume concentration is usually used when the solute is alcohol. The calculation is similar to percent by mass except volumes are used.

$$\% \text{ (volume)} = \frac{\text{volume of solute}}{\text{volume of solution}} \times 100\%$$

SAMPLE PROBLEM

What is the percent by volume concentration of 10 mL of ethanol dissolved in 50. mL of water?

Solution:

1. Find the total volume:

$$10. \text{ mL} + 50. \text{ mL} = 60. \text{ mL}$$

2. Multiply the ratio of solute to solvent by 100%

$$\frac{10. \text{mL}}{60. \text{mL}} \times 100\% = \textbf{17\% by volume}$$

Parts per Million. Parts per million (ppm) is often used when only small amounts of solute are dissolved. It is the number of milligrams (10^{-3}g) of solute per liter of solvent.

$$\text{ppm} = \frac{\text{mg}}{\text{L}}$$

SAMPLE PROBLEM

Find the concentration in ppm if 0.10 g of sugar is dissolved in 500. mL of water.

Solution:

   1. Change grams of solute into mg:

$$0.10 \text{ g} \times \frac{1000 \text{ mg}}{1.0 \text{ g}} = 100 \text{ mg}$$

   2. Change 500 mL to liters:

$$500 \text{ mL} \times \frac{1.00 \text{ L}}{1000 \text{ mL}} = .500 \text{ L}$$

   3. Calculate ppm:

$$\text{ppm} = \frac{100. \text{ mg}}{.500 \text{ L}} = \textbf{200. ppm}$$

Boiling Point Elevation and Freezing Point Depression

While pure substances melt and boil at characteristic, constant temperatures, the melting and boiling points of solutions depend on their concentrations. The greater the number of solute molecules, or ions, in a kilogram of solution, the higher the boiling point, the lower the vapor pressure, and the lower the freezing point.

QUESTIONS

Answer the following questions using Tables C, G, F, and H of the *Reference Tables for Physical Setting/Chemistry*.

1. If 60.g of KNO_3 (s) is dissolved in 100. mL of water at 50.°C, the solution will be

   (1) saturated   (2) unsaturated   (3) supersaturated   (4) dilute

2. At what temperature will 50. grams of NH_4Cl just saturate 100. grams of water?

   (1) 28°C   (2) 33°C   (3) 48°C   (4) 56°C

3. How many mg is 0.020g?

   (1) 20.   (2) 2.0   (3) 200.   (4) 2000.

4. How many mL of ethanol would be in 50. mL of an aqueous solution if the concentration was 10.% by volume?

   (1) 1.0   (2) 2.0   (3) 5.0   (4) 10.

5. How many grams of an aqueous solution containing 5.0 g of NaCl has a concentration of 15% by mass?

   (1) 0.75 g   (2) 0.33 g   (3) 15 g   (4) 33 g

6. How many grams of solute must be in 2.0 L of water to have a concentration of 8.0 ppm?

   (1) 0.004 g   (2) 0.008 g   (3) 4.0 g   (4) 0.016 g

7. How many grams of ammonium chloride (gram formula mass = 53.5 g) are contained in 0.500 L of a 2.00 M solution?

   (1) 10.0 g   (2) 26.5 g   (3) 53.5 g   (4) 107 g

8. What is the molarity of an H_2SO_4 solution if 0.25 liter of the solution contains 0.75 mole of H_2SO_4?

   (1) 0.33 M   (2) 0.75 M   (3) 3.0 M   (4) 6.0 M

9. What is the total number of grams of NaOH (formula mass = 40.) needed to make 1.0 liter of a 0.20 M solution?

   (1) 20. g   (2) 2.0 g   (3) 80. g   (4) 8.0 g

10. As additional $KNO_3(s)$ is added to a saturated solution of KNO_3 at constant temperature, the concentration of the solution

    (1) decreases   (2) increases   (3) remains the same

11. If 0.50 liter of a 12-molar solution is diluted to 1.0 liter, the molarity of the new solution is

(1) 2.4   (2) 6.0   (3) 12   (4) 24

12. A 1 kilogram sample of water will have the highest freezing point when it contains

(1) 1×10^{17} dissolved particles   (2) 1×10^{19} dissolved particles

(3) 1×10^{21} dissolved particles   (4) 1×10^{23} dissolved particles

13. Compared to the normal freezing point and boiling point of water, a 1-Molar solution of sugar in water will have a

(1) higher freezing point and a lower boiling point
(2) higher freezing point and a higher boiling point
(3) lower freezing point and a lower boiling point
(4) lower freezing point and a higher boiling point

14. Which ratio of solute-to-solvent could be used to prepare a solution with the highest boiling point?

(1) 1 g of NaCl dissolved per 100 g of water
(2) 1 g of NaCl dissolved per 1000 g of water
(3) 1 g of $C_{12}H_{22}O_{11}$ dissolved per 100 g of water
(4) 1 g of $C_{12}H_{22}O_{11}$ dissolved per 1000 g of water

PRACTICE FOR CONSTRUCTED RESPONSE

1. Sketch in the appropriate boxes below a model of molecules, ⬭ , showing their different arrangement as a solid, a liquid, and a gas. (3 points)

solid	liquid	gas

2. The saying "Like dissolves like" means polar solvents dissolve polar solutes better than non-polar solvents. Show, with a diagram, why this would be true. Explain your diagram. (3 points)

polar solvent

non-polar solvent

_____

_____

3. A constant amount of gas is sealed in an expandable container. The volume is read as the temperature and pressure vary, yielding the data on the right.

a. Plot the volume as the pressure changes at constant temperature. (2 points)

$V_{(mL)}$	$T_{(°C)}$	$P_{(atm)}$
800.	20.	0.730
827	30	0.730
855	40	0.730
882	50	0.730
870	50	0.740
858	50	0.750
847	50	0.760
861	60	0.770
876	70	0.780
891	80	0.790

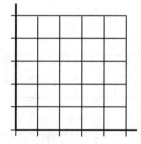

b. Plot the volume as the Kelvin temperature changes at constant pressure. (2 points)

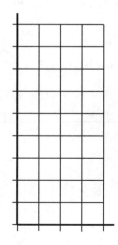

c. What relationship exists between pressure and volume of a gas at constant temperature? (1 point)

_____

_____

d. What relationship exists between Kelvin temperature and volume of a gas at constant pressure? (1 point)

_____

_____

e. Calculate what the volume would be at 100.°C and 0.700 atm. (2 points)

_____

4. 100. grams of liquid X is cooled in a water bath yielding the following data:

time (min)	0.0	1.0	2.0	3.0	4.0	5.0	6.0	7.0	8.0
Temperature (°C)	77	71	65	59	53	53	53	53	53

a. Plot the graph on the axes provided. (2 points)

b. At what time does X begin to freeze? (1 point)

_____

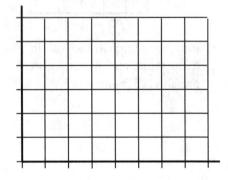

   c. Calculate how much heat is lost between the times of 0.0 and 4.0 minutes. (The specific heat capacity of X is 1.00 J/g·K) Show your work. (2 points)

   _____

   d. During the time interval 4.0 to 8.0 minutes, does X _____
   (1 point)

      (1) gain heat   (2) lose heat   (3) keep the same heat

   e. Is X pure or a mixture? Support your answer from your graph or the data.
   (2 points)

   _____

   _____

   f. What changes occur in the graph if 5 grams of sugar are added and dissolved in X?
   (1 point)

   _____

   _____

5. 5.00 g of $CaCl_2$ is dissolved in enough water to make 2.00 liters of solution. Calculate the concentration in

   a. Molarity
   (2 points)

   _____

   b. parts per million
   (2 points)

   _____

CHAPTER 6
KINETICS AND EQUILIBRIUM

KINETICS

Chemical kinetics deals with:

• The rates of chemical reactions
• The pathway by which the reactions occur

A **reaction rate** depends on several variables, including:

• the nature of the reactants
• the concentration of the reactants
• surface area
• presence of a catalyst
• the temperature of the system

Rate of reaction is the change in concentration of a given substance (reactant or product) per unit time. Reaction rate is measured in terms of moles of reactant consumed (or product formed) per liter per second.

Very few chemical reactions occur as directly as an equation for that reaction indicates. In many cases, an equation represents a *net* reaction, which is a summation of several intermediate reactions. These intermediate equations represent the actual *reaction pathway* of the overall reaction.

Role of Energy in Reactions

In order for a chemical reaction to begin, energy is needed. As a result of the reaction, energy may be released or absorbed by the system. A complete picture of the energy factors in a chemical reaction is best illustrated by a potential energy diagram in which the potential energy of the substances involved in the reaction is plotted against a time sequence.

Activation energy. Activation energy is the minimum energy needed to cause a reaction to begin.

Heat of reaction. Heat of reaction is the difference between the potential energy of the products and that of the reactants. The potential energy of molecules was once referred to as "heat content." This obsolete and misleading term accounts for the use of the symbol H in the equation:

$$\Delta H = H_{products} - H_{reactants}$$

The Greek letter Δ, called *delta,* denotes a "difference." ΔH is the symbol for Heat of Reaction. A number of ΔH's for reactions are listed on Table I of the *Reference Tables for Physical Setting/Chemistry.*

Exothermic reactions are those that release energy. In such reactions, the potential energy of the products is lower than that of the reactants. In exothermic reactions, the sign of ΔH is negative. Consider the reaction:

$$H_2(g) + \tfrac{1}{2} O_2(g) \rightarrow H_2O(g) + 241.8 \text{ kJ}$$

The reaction releases heat. It is exothermic. ΔH for this reaction is -241.8 kJ/mole H_2O.

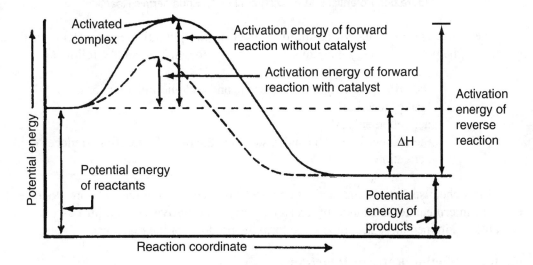

Figure 6-1. Potential Energy Diagram For An Exothermic Reaction

Endothermic reactions are those that absorb energy. In such reactions, the potential energy of the products is greater than that of the reactants. In endothermic reactions, the sign of ΔH is positive. Consider the reaction:

$$\tfrac{1}{2}H_2(g) + \tfrac{1}{2}I_2(g) + 26.5 \text{ kJ} \rightarrow HI(g)$$

This reaction is endothermic, absorbing energy.

$$\Delta H \text{ is } + 26.5 \text{ kJ/mole HI}$$

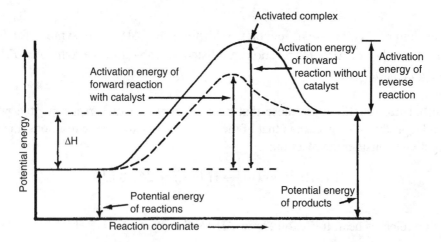

Figure 6-2. Potential Energy Diagram For An Endothermic Reaction

Heat of reaction is shown as the difference in the potential energy (P.E.) in these reactions. The potential energy diagrams make it easier to demonstrate the following:

- the difference between exothermic and endothermic reactions.
- heat of reaction
- activation energy
- the manner in which catalysts change the rate of a reaction (reduces activation energy)

Most chemists believe that activation energy is used to bring the reactants together to form an intermediate substance of greater energy. The **activated complex** is extremely unstable and immediately decomposes to form the products of the reaction.

Factors Affecting Rates of Reaction

It is believed that in order for particles (atoms, molecules, or ions) to react, they must collide with each other. Calculations have shown however, that when the number of collisions is increased, the rate of reaction is *not* increased proportionately. This discovery has led to the conclusion that not all collisions are *effective* in bringing about reaction. In order to be effective, collisions must have (1) enough energy and (2) the proper orientation of the colliding particles.

The effectiveness of collisions is determined by several factors:

Nature of the reactants. Chemical reactions occur by the breaking and the rearranging of existing bonds and by the forming of new bonds. The fewer the rearrangements that occur, the faster the reaction takes place. Reactions involving ions in aqueous solution are usually extremely rapid, probably because there are no bonds to be broken. Reactions between molecules require the breaking of bonds. At room temperature, reactions such as the one between hydrogen (H_2) and oxygen (O_2) are usually very slow.

Concentration. An increase in the concentration of a reactant increases the frequency of collisions. The result is an increase in reaction rate. It is customary in chemical kinetics to express concentrations in moles per liter (molarity). Brackets around a formula or symbol are used to indicate concentration. [Cl⁻] means the concentration of the chloride ion in molarity.

In reactions involving gases in closed systems, *increased pressure* results in increased concentration. The net result is an increase in the rate of reaction.

Temperature. Increasing the temperature of the system always increases the rate of reaction. Since temperature represents the average kinetic energy of the particles, the effectiveness of the collisions, as well as the number of collisions is increased.

Surface Area. As one might expect, increasing the surface area of reactants provides greater opportunity for collisions to occur. This can easily be demonstrated in the reaction between zinc metal and hydrochloric acid to produce hydrogen gas. Reducing the size of the zinc particles increases the surface area of the zinc. Evidence of the increase in reaction rate is provided by the increased bubbling resulting from hydrogen formation.

Catalysts. Catalysts, by definition, are substances that increase the rate of reaction and seem to remain unchanged. This increased rate is brought about through a change to an alternate reaction pathway so that less activation energy is required. Catalyzed reactions take place more rapidly than uncatalyzed reactions because of the reduction in activation energy. The potential energy of the reactants and the potential energy of the products remain the same. Therefore, the heat of reaction is the same, whether catalyzed or not.

QUESTIONS

Answer the following questions using Table I of the *Reference Tables for Physical Setting/Chemistry.*

1. An increase in the temperature increases the rate of a chemical reaction because the collisions in this reaction increase in

(1) number only   (2) effectiveness, only
(3) both number and effectiveness   (4) neither number nor effectiveness

2. In a chemical reaction, the difference between the potential energy of the products and the potential energy of the reactants is the

(1) heat of reaction   (2) heat of fusion   (3) free energy   (4) activation energy

Base your answers to questions 3 and 4 on the potential energy diagram below

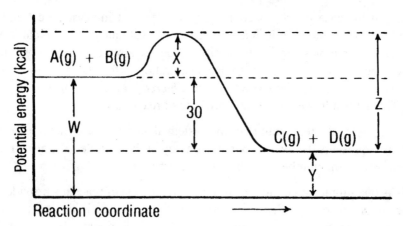

3. The potential energy of the activated complex is equal to the sum of

(1) X + Y   (2) X + W   (3) X + Y+ W   (4) X + W + Z

4. The reaction $A(g) + B(g) \rightarrow C(g) + D(g) + 30$ kcal has a forward activation energy of 20 kcal. What is the activation energy for the reverse reaction?

(1) 10 kcal   (2) 20 kcal   (3) 30 kcal   (4) 50 kcal

5. The graph below represents a chemical reaction.

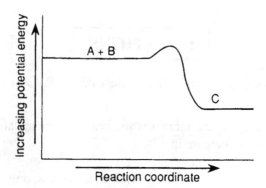

This reaction is best described as

(1) endothermic, because energy is absorbed
(2) endothermic, because energy is released
(3) exothermic, because energy is absorbed
(4) exothermic, because energy is released

6. The potential energy diagram shown below represents the reaction $R + S +$ energy $\rightarrow T$.

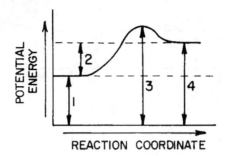

Which numbered interval represents the potential energy of the product T?

(1) 1   (2) 2   (3) 3   (4) 4

7. A potential energy diagram is shown below.

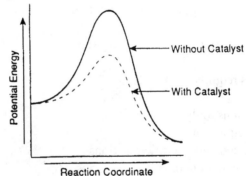

Which reaction would have the *lowest* activation energy?

(1) the forward catalyzed reaction   (2) the forward uncatalyzed reaction
(3) the reverse catalyzed reaction   (4) the reverse uncatalyzed reaction

8. The potential energy diagram of a chemical reaction is shown below.

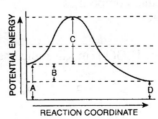

Which letter in the diagram represents the heat of reaction (ΔH)?

(1) A   (2) B   (3) C   (4) D

9. The potential energy diagram below represents the reaction $2KClO_3 \rightarrow 2KCl + 3O_2$

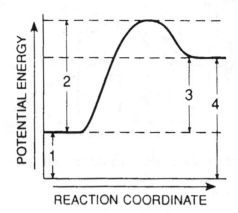

Which numbered interval on the diagram would change when a catalyst is added?

(1) 1   (2) 2   (3) 3   (4) 4

10. Activation energy is required to initiate

(1) exothermic reactions, only
(2) endothermic reactions, only
(3) both exothermic and endothermic reactions
(4) neither exothermic nor endothermic reactions

11. Given the reaction: A + B → C + D
    The reaction will most likely occur at the greatest rate if A and B represent

(1) nonpolar molecular compounds in the solid phase
(2) ionic compounds in the solid phase
(3) solutions of nonpolar molecular compounds
(4) solutions of ionic compounds

12. What is the heat of formation of $H_2O(1)$, in kiloJoules per mole, at 1 atmosphere and 298 K?
    (1) − 285.8   (2) − 571.6   (3) − 2259   (4) − 1572

13. According to the *Reference Tables for Physical Setting/Chemistry* Table I ,the dissolving of which salt is accompanied by the release of energy?

(1) LiBr   (2) NH_4Cl (3) NaCl   (4) KNO_3

14. Which change would most likely increase the rate of a chemical reaction?

(1) decreasing a reactant's concentration
(2) decreasing a reactant's surface area
(3) cooling the reaction mixture
(4) adding a catalyst to the reaction mixture

EQUILIBRIUM

Since most reactions proceed in both directions, it is necessary to study the rates of both the forward and reverse reactions. When the forward and reverse rates are equal in such *reversible systems,* the system is said to be in a state of **dynamic equilibrium**. The word dynamic suggests motion. Dynamic equilibrium refers to a state of balance between two opposing activities in which the concentration of both reactants and products remain *constant.* These concentrations are not necessarily equal, however. A state of equilibrium may exist in which quantities of reactants and products are quite different. Furthermore, the reversibility of reactions permits the attainment of equilibrium from either the forward or the reverse reaction.

Phase Equilibrium

When substances change phase, the changes are reversible. In closed systems, a state of equilibrium between phases may be achieved. Whenever liquids or solids are confined in a closed container, the condition will be reached in which there are enough particles in the gas (vapor) phase to cause the rate of return to the original phase to be equal to the rate of escape. This condition will result in a vapor pressure that is temperature dependent and characteristic for the particular solid or liquid involved.

Solution Equilibrium

Gases in liquids. Liquids containing dissolved gases in closed systems will reach equilibrium not only between the liquid and its gas phase but also between the dissolved gas and the undissolved gas above the liquid. This equilibrium is affected by temperature and pressure, since increased temperature and/or reduced pressure reduces the solubility of gases in liquids. Low temperatures and high pressures favor solution of gases in liquids. Carbonated beverages, therefore, maintain their carbonation best when tightly covered and chilled.

Solids in liquids. When a solution of a solid in a liquid contains all the solid that will dissolve at existing conditions, it is said to be a saturated solution. If additional solute is added, it will fall to the bottom of the container. A condition of equilibrium may be reached between the dissolved and undissolved solute. In this condition the processes of dissolving and crystallizing of the solute occur at equal rates.

Solubility. The concentration of the solute in the saturated solution described above is referred to as the **solubility** of the solute in that liquid. To review the vocabulary of solutions:

- The solute is the substance dissolved.
- The solvent is the substance in which the solute is dissolved.
- The solution is the homogeneous mixture that results.

Chemical Equilibrium

In any chemical reaction, equilibrium is recognized when observable changes, such as color, temperature, pressure etc. (known as macroscopic changes) no longer occur. At this point the forward and reverse reactions are occurring at equal rates.

Le Chatelier's Principle. Chemists refer to changes in concentration, pressure, or temperature as **applied stresses**. Le Chatelier's Principle is a generalization that describes what happens to a system subjected to such stresses. When a stress is applied to a system in equilibrium, the reaction will try to shift in the direction that will relieve the stress. Equilibrium will then be re-established at a different point.

This principle, and how it operates, is well demonstrated in the *Haber process* for the manufacture of ammonia (NH_3) from nitrogen (N_2) and hydrogen (H_2). The equation for the reaction is:

$$N_2\,(g) + 3H_2\,(g) \rightleftarrows 2NH_3\,(g) + 91.8\ kJ$$

The equation for the reaction provides the following information:

1. The reaction is *exothermic,* since there is a release of heat for the NH_3 produced. If the reaction followed the rules of stoichiometry, 4 moles of reactant would yield 2 moles of product, resulting in a reduced volume.

2. Since the reactants and the products are gases, they are subject to both pressure and temperature changes.

a. *Effect of concentration.* Increasing the concentration of one substance in a reaction at equilibrium will cause the reaction to proceed in the direction that will use up the increase. In time, a new equilibrium will be established. In the Haber process, increasing the concentration of nitrogen or hydrogen will increase the rate of ammonia production. Of course, if the system remains closed, the increased concentration of ammonia will cause an increase in the reverse reaction until a new equilibrium is reached. If ammonia is removed, however, the resulting decrease in concentration of product will further the forward reaction so as to increase the output of ammonia.

Products may be removed from a reaction by
  (1) the formation of a gas
  (2) the formation of an insoluble product (precipitate) by the reaction of
      dissolved reactants, or
  (3) the formations of a non-electrolyte such as water.

Such reactions are *reactions that go to completion.*

b. *Effect of pressure.* Changes in pressure affect only the gaseous components in an equilibrium system. Increased pressure favors the reaction that results in a smaller volume caused by fewer number of gaseous molecules. Increasing the pressure in the Haber process, therefore, will favor the forward reaction, since the production of ammonia results in fewer molecules and reduced volume. This reduction in volume relieves the stress caused by the increased pressure. In reactions in which the volume of reactant(s) is equal to the volume of product(s), changes in pressure have no effect and therefore the equilibrium will not shift in either direction.

c. *Effect of temperature.* When the temperature of a system in equilibrium is raised, the equilibrium shifts in the direction that will absorb that heat. An endothermic reaction is favored. A decrease in temperature of a system in equilibrium favors an exothermic reaction. For the Haber process, an increase in temperature causes the reaction to shift to the reactants favoring the decomposition of ammonia.

d. *Effect of catalysts.* The addition of a catalyst to a system may cause equilibrium to be reached more rapidly. There will be *no net change* in the equilibrium concentration, because catalysts increase rates of forward and reverse reactions equally.

Spontaneous Reactions

Whether reactions proceed or not, seems to depend on the balance between two basic natural tendencies:

1. the drive toward greater stability (reduced potential energy)

2. the drive toward less organization (increased entropy)

Energy changes. Systems in conditions of great energy tend to change to conditions of low energy. Given ΔH as the symbol for energy change in chemical systems, this tendency favors reactions in which ΔH is negative; that is, *exothermic* reactions.

Entropy changes. Systems tend to change from conditions of great order to conditions of low order. This tendency is interpreted as a change from low entropy to high entropy. The term randomness is frequently used to describe entropy. The more random the system, the higher the entropy. High entropy is favored by increased temperature.

The symbol for entropy is S. A positive ΔS indicates a phase change of increasing randomness:

$$\text{solid} \text{-----}>\text{liquid}\text{----->}\text{gas} \qquad \Delta S = (+)$$

At constant temperature, substances in a system tend to change phase so that, in its final state, the system has higher randomness (entropy) than it had in its initial state.

Gas molecules have greater entropy than liquid molecules because the gas molecules have no volume of their own. Molecules of liquids have greater entropy than molecules of solids because the liquids' molecules are not organized into a definite shape.

QUESTIONS

1. A solution in which equilibrium exists between undissolved and dissolved solute is always

   (1) saturated   (2) unsaturated   (3) dilute   (4) concentrated

2. A flask at 25°C is partially filled with water and stoppered. After a period of time the water level remained constant. Which relationship best explains this observation?

   (1) The rate of condensation exceeds the rate of evaporation.
   (2) The rates of condensation and evaporation are both zero.
   (3) The rate of evaporation exceeds the rate of condensation.
   (4) The rate of evaporation equals the rate of condensation.

3. Given the equation:

$$H_2(g) + I_2(g) \rightleftarrows 2HI(g)$$

   Which statement is always true when this reaction has reached chemical equilibrium?

   (1) $[H_2] \times [I_2] > [HI]$.   (2) $[H_2] \times [I_2] < [HI]$.
   (3) $[H_2]$, $[I_2]$, and $[HI]$ are all equal.   (4) $[H_2]$, $[I_2]$, and $[HI]$ remain constant.

4. Given the reaction:

$$HC_2H_3O_2(aq) + H_2O \rightleftarrows H_3O^+(aq) + C_2H_3O_2^-(aq)$$

   When the reaction reaches a state of equilibrium, the concentrations of the reactants

   (1) are less than the concentrations of the products
   (2) are equal to the concentrations of the products
   (3) begin decreasing
   (4) become constant

5. Given the reaction at equilibrium:

$$CO_3^{2-}(aq) + H_2O(\ell) \rightleftarrows HCO_3^-(aq) + OH^-(aq)$$

Which statement is always true?

(1) $[CO_3^{2-}]$ is less than $[OH^-]$.
(2) $[CO_3^{2-}]$ is less than $[HCO_3^-]$.
(3) The rate of the forward reaction is less than the rate of the reverse reaction.
(4) The rate of the forward reaction equals the rate of the reverse reaction.

6. Given the system $CO_2(s) \rightleftarrows CO_2(g)$ at equilibrium. As the pressure increases at constant temperature, the amount of $CO_2(g)$ will

(1) decrease   (2) increase   (3) remain the same

7. Which system at equilibrium will shift to the right when the pressure is increased?

(1) $NaCl(s) \overset{H_2O}{\rightleftarrows} Na^+(aq)\ Cl^-(aq)$

(2) $C_2H_5OH(\ell) \overset{H_2O}{\rightleftarrows} C_2H_5OH(aq)$

(3) $NH_3(g) \overset{H_2O}{\rightleftarrows} NH_3(aq)$

(4) $C_6H_{12}O_6(s) \overset{H_2O}{\rightleftarrows} C_6H_{12}O_6(aq)$

8. Given the reaction at equilibrium:

$$2SO_2(g) + O_2(g) \rightleftarrows 2SO_3(g) + heat$$

The rate of the forward reaction can be increased by adding more SO_2 because the

(1) temperature will increase
(2) number of molecular collisions between reactants will increase
(3) reaction will shift to the left
(4) forward reaction is endothermic

9. Given the reaction at equilibrium:

$$CO(g) + \tfrac{1}{2}O_2(g) \rightleftarrows CO_2(g) + 67.7 \text{ kcal}$$

As the temperature increases, the rate of the forward reaction

(1) decreases   (2) increases   (3) remains the same

10. The diagram below shows a bottle containing $NH_3(g)$ dissolved in water. How can the equilibrium $NH_3(g) \rightleftarrows NH_3(aq)$ be reached?

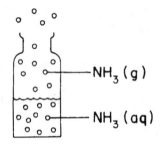

(1) Add more water   (2) Add more $NH_3(g)$
(3) Cool the contents   (4) Stopper the bottle

11. A sample of H_2O (ℓ) at 20° C is in equilibrium with its vapor in a sealed container. When the temperature increases to 25°C, the entropy of system will

(1) decrease   (2) increase   (3) remain the same

12. Above 0°C, ice changes spontaneously to water according to the following equation: $H_2O(s)$ + heat $\rightarrow$ $H_2O(\ell)$. The changes in $H_2O(s)$ involve

(1) an absorption of heat and a decrease in entropy
(2) a release of heat and a decrease in entropy
(3) an absorption of heat and an increase in entropy
(4) a release of heat and an increase in entropy

13. As 1 gram of H_2O (g) changes to 1 gram of H_2O (ℓ), the entropy of the system

(1) decreases   (2) increases   (3) remains the same

14. Which reaction results in an increase in entropy?

(1) $2H_2O(\ell) \rightarrow 2H_2(g) + O_2(g)$ (2) $2H_2(g) + O_2(g) \rightarrow 2H_2O(\ell)$
(3) $N_2(g) + 3H_2(g) \rightarrow 2NH_3(g)$ (4) $2CO(g) + O_2(g) \rightarrow 2CO_2(g)$

15. A 1 gram sample of a substance has the greatest entropy when it is in the

(1) solid state   (2) liquid state   (3) crystalline state   (4) gaseous state

16. The diagram below shows a system of gases with the valve closed.

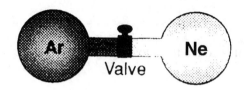

As the valve is opened, the entropy of the gaseous system

(1) decreases   (2) increases   (3) remains the same

PRACTICE FOR CONSTRUCTED RESPONSE

1. For the reaction A $\rightarrow$ B the following information is given:

$$\Delta H = -10 \text{ kJ/mol}$$
$$\text{potential energy (H) of A} = 15 \text{ kJ/mol}$$
$$\text{activation energy} = 7 \text{ kJ/mol}$$

a. Draw the potential energy graph of the reaction A $\rightarrow$ B on the axes provided. (2 points)

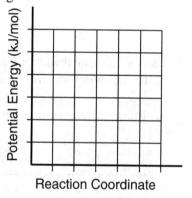

b. What is the potential energy of B? (1 point) _____

c. Is the reaction endothermic or exothermic? (1 point) _____

d. What part of the graph is affected by a catalyst? (1 point) _____

e. What is the activation energy for the reaction B $\rightarrow$ A? _____ (1 point)

2. Which of the following systems are at dynamic equilibrium, and which are not? Provide the reason for your answer.

   a. Solid sodium chloride is added to water in a beaker. It stops dissolving with some solid left on the bottom of the beaker. (2 points)

   _____

   b. A drop of alcohol is added to a closed flask. The alcohol completely evaporates. (2 points)

   _____

   c. 50 mL of alcohol is placed in a 100 mL beaker. (2 points)

   _____

   d. Sodium chromate (yellow) is added to water. The solution continues to become darker yellow. (2 points)

   _____

3. Use the collision theory of reactions and the kinetic molecular theory to explain how each of the following actions will increase a reaction rate.

   a. Raise the temperature. (1 point)

   _____

   b. Crush a solid before reacting. (1point)

   _____

   c. Increase the concentration of a reactant. (1 point)

   _____

4. Two liquids, A (ℓ) and X (ℓ) and two gases B (g) and Y (g) are involved in the following four reactions. The heats of reaction are given for two of the reactions.

1. A (ℓ) $\rightarrow$ B (g) $\Delta H = + 5\,kJ/mol$
2. B (g) $\rightarrow$ A (ℓ)
3. X (ℓ) $\rightarrow$ Y (g) $\Delta H = - 10\,kJ/mol$
4. Y (g) $\rightarrow$ X (ℓ)

a. Which of these reactions will definitely react spontaneously? (2 points)

_____

_____

b. Which of these reactions will definitely *not* react spontaneously? (2 points)

_____

_____

5. Over 36 million kilograms of sulfuric acid are made each year, the most of any chemical. It is used in the manufacture of fertilizers, batteries, detergents, and plastics. To make it, SO_3(g) is bubbled into water. But first, SO_2(g) must be made by burning sulfur, and then SO_3(g) is made by this reaction:

$$2\,SO_2(g) + O_2(g) = 2SO_3(g) + 200kJ$$

Describe three changes in this equilibrium that would promote the production of more SO_3(g). (3 points)

1._____

2._____

3._____

CHAPTER 7
ORGANIC CHEMISTRY

Organic chemistry is the name given to the study of carbon and carbon compounds. Carbon atoms have a very strong tendency to form four covalent bonds. Because of this tendency, carbon atoms produce an enormous variety of compounds. They form bonds not only with atoms of other elements, they also form bonds with other carbon atoms. The number of carbon compounds (or organic compounds) is far greater than the number of inorganic compounds-perhaps as much as thirty times greater.

Living things contain many carbon compounds. Products of living things, such as petroleum, wood, and coal supply the raw materials from which most organic chemicals are obtained.

BONDING

Bonding

Carbon has an electronegativity of 2.6, about in the middle of the electronegativity range of 0.7 to 4.0. Since other elements are not likely to attract electrons much more or less than carbon, electrons are shared, and covalent bonds result. A carbon atom has four valence electrons, therefore it shares four pairs of electrons (has four bonds) with other atoms. The four bonds spread out to the corners of a tetrahedron around the carbon atom. (See p. 51)

When carbon atoms form bonds with other carbon atoms, they can share one, two, or even three pairs of electrons, the latter cases resulting in double and triple bonds. Because the bonding is covalent, carbon compounds are molecular.

Structural Formulas

The importance of the three-dimensional nature of organic compounds has led to the widespread use of models to illustrate the bonding within organic molecules. Another device for representing the arrangement of atoms in an organic compound is the **structural formula.** In a structural formula, each covalent bond is represented by a short, straight line between atoms. Consider the structural formula of methane CH_4 compared to its electron dot formula:

$$
\begin{array}{c}
H \\
| \\
H-C-H \\
| \\
H
\end{array}
\qquad
\begin{array}{c}
H \\
\cdot x \\
H \mathbin{:} C \mathbin{:} H \\
x \cdot \\
H
\end{array}
$$

Methane   Methane

This structural formula provides an approximation of the shape of the molecule. A ball-and-stick model can be used to show, in three dimensions, the actual tetrahedral arrangement of the bonds in methane.

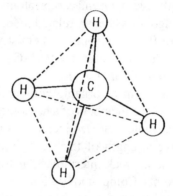

Figure 9-1. Ball-and Stick Model.

Isomers

Many organic compounds have the same molecular formula—that is, the same atoms in the same molar ratios—but display vastly different properties. Investigations have revealed that while molecular formulas may be the same, the arrangements of the atoms are quite different. Compounds that have the same molecular formula but different structural formulas are called **isomers**. Propanone and propanal are examples of isomers. Both have the molecular formula C_3H_6O. Structural formulas of these two compounds show the different arrangements of the atoms.

$$
\begin{array}{ccc}
H & O & H \\
| & \| & | \\
H-C-C-C-H \\
| & & | \\
H & & H
\end{array}
\qquad\qquad
\begin{array}{ccc}
H & H & O \\
| & | & \| \\
H-C-C-C-H \\
| & | \\
H & H
\end{array}
$$

Propanone   proponal

Structural formulas are somewhat cumbersome and require considerable space. An intermediate device, between molecular and structural formulas, is often used to represent organic compounds on a single typed line. These formulas show the atoms in the "order" in which they are arranged in the molecule. For example, the formula for propanone is written CH_3COCH_3; that for propanal is written CH_3CH_2CHO. These formulas are known as condensed structural formulas.

HYDROCARBONS

Compounds containing only hydrogen and carbon atoms are called **hydrocarbons.** Many organic compounds can be thought of as being related to, or derived from, hydrocarbons.

The International Union of Pure and Applied Chemistry (IUPAC) has devised a system for naming organic compounds. In many cases, the IUPAC names have been derived from the hydrocarbons to which the compounds are related.

A homologous series is a group of organic compounds with similar properties and related structures. The formulas of the members of a homologous series differ from each other by some common increment. As molecular size increases among members of a homologous group, so too does the relative strength of the weak intermolecular forces holding the molecules together. The increase in weak intermolecular forces results in increased boiling points and freezing (melting) points. Compounds with low molecular masses tend to be gases. Those with the highest molecular masses tend to be solids under normal conditions of temperature and pressure.

Saturated and Unsaturated Compounds

Saturated compounds are organic compounds in which all the carbon-to-carbon bonds are formed by the sharing of single pairs of electrons. In other words, all carbon-to-carbon bonds are single bonds.

Unsaturated compounds are those in which at least one of the carbon-to-carbon bonds is a double or triple bond. A double bond is formed from two shared pairs of electrons. A triple bond contains three shared pairs of electrons.

Alkanes

The **alkane** series—also called the methane series or the paraffin series—is the group of saturated hydrocarbons with the general formula C_nH_{2n+2}. The names of the compounds all end in *-ane*. The first part of the name is related to the number of carbons: *meth-* for 1 carbon, *eth-* for 2 carbons, *prop-* for 3 carbons, *but-* for 4 carbons and *pent-* for 5 carbons. See the *Reference Tables for Chemistry* Table P for 6-10 carbons.

TABLE 7-1. FIRST FIVE MEMBERS OF THE ALKANE SERIES

Hydrocarbon	Molecular formula	Structural formula
methane	CH_4	H | H – C – H | H
ethane	C_2H_6	H H | | H – C – C – H | | H H
propane	C_3H_8	H H H | | | H – C – C – C – H | | | H H H
butane	C_4H_{10}	H H H H | | | | H – C – C – C – C – H | | | | H H H H
pentane	C_5H_{12}	H H H H H | | | | | H – C – C – C – C – C – H | | | | | H H H H H

The alkane series begin to show isomerism with butane, C_4H_{10}.

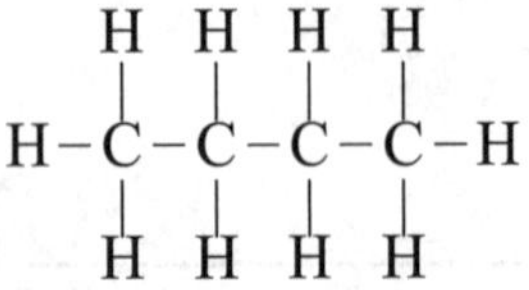

normal or n-butane

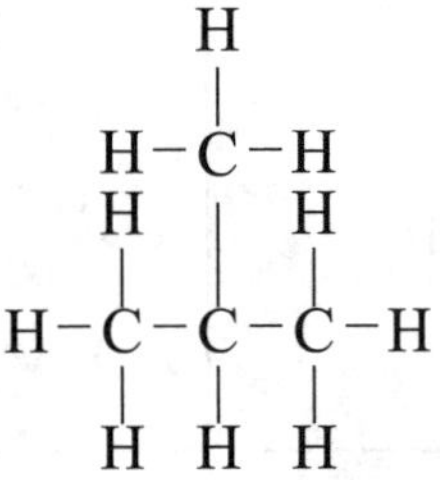

methyl propane

Alkenes

The members of the **alkene** series are unsaturated hydrocarbons containing one double bond. Alkenes have the general formula C_nH_{2n} The alkenes are named by changing the ending of the corresponding alkane from -*ane* to -*ene* The first four alkenes are ethene, propene, butene, and pentene.

TABLE 7-2. FIRST FOUR MEMBERS OF THE ALKENE SERIES

Hydrocarbon	Molecular formula	Structural formula
ethene	C_2H_4	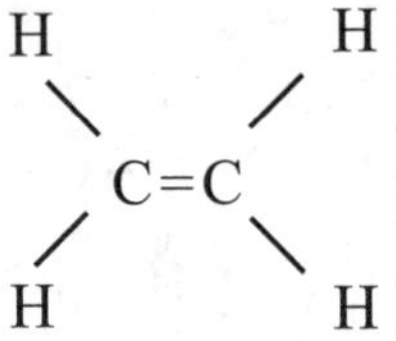
propene	C_3H_6	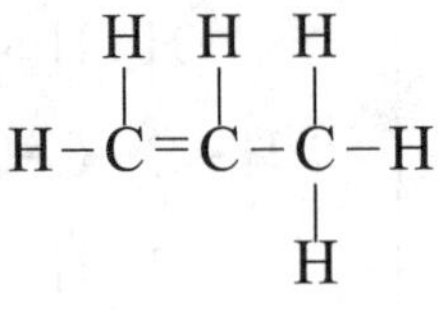

| butene | C_4H_8 | 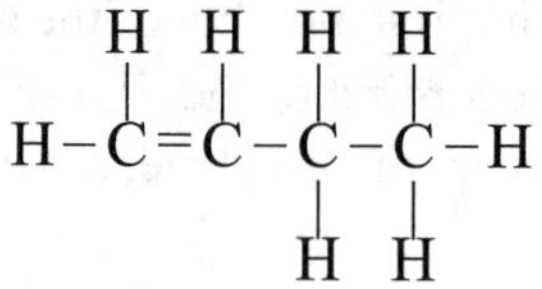 |
| pentene | C_5H_{10} | 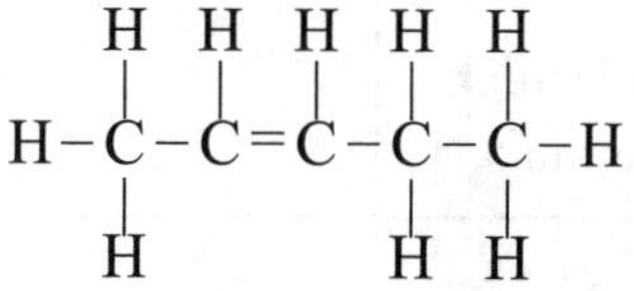 |

The position of the double bond in long chains can vary. These are also structural isomers. The longest chain of carbons are numbered, starting closest to the double bond. The lower number of the carbons with the double bond precedes the name of the hydrocarbon.

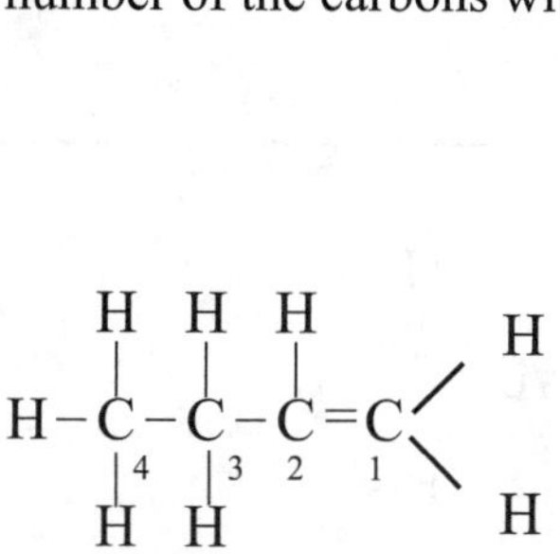

1-butene

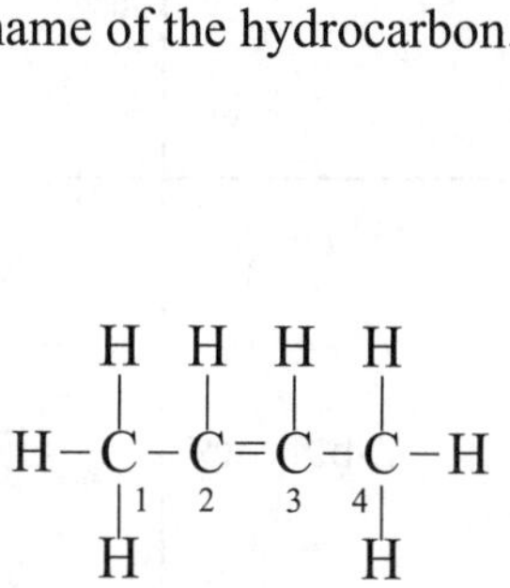

2-butene

Alkynes

Members of the **alkyne** series are unsaturated hydrocarbons containing one triple bond. Their general formula is C_nH_{2n-2}. Members are named from the corresponding alkane by changing the –*ane* to –*yne*. The first member of the group, $H-C\equiv C-H$, is best known by its common name, *acetylene*. According to the IUPAC system, the name is *ethyne*. The series, accordingly, is called the ethyne, or acetylene, series.

TABLE 7-3. FIRST FOUR MEMBERS OF THE ALKYNE SERIES

Hydrocarbon	Molecular formula	Structural formula
ethyne	C_2H_2	$H-C\equiv C-H$
propyne	C_3H_4	$H-C\equiv C-\overset{\displaystyle H}{\underset{\displaystyle H}{C}}-H$
2-butyne	C_4H_6	$H-\overset{\displaystyle H}{\underset{\displaystyle H}{C}}-C\equiv C-\overset{\displaystyle H}{\underset{\displaystyle H}{C}}-H$
2-pentyne	C_5H_8	$H-\overset{\displaystyle H}{\underset{\displaystyle H}{C}}-C\equiv C-\overset{\displaystyle H}{\underset{\displaystyle H}{C}}-\overset{\displaystyle H}{\underset{\displaystyle H}{C}}-H$

QUESTIONS

Answer the following questions using Tables P and Q of the *Reference Tables for Physical Setting/Chemistry.*

1. All organic compounds must contain the element

   (1) hydrogen   (2) nitrogen   (3) carbon   (4) oxygen

2. The four single bonds of a carbon atom are spatially directed toward the corners of a regular

   (1) triangle   (2) rectangle   (3) square   (4) tetrahedron

3. Which pair of compounds are isomers?

   (1) C_6H_6 and C_6H_{12} (2) C_2H_4 and C_2H_6
   (3) CH_3CH_2OH and CH_3COOH (4) CH_3CH_2OH and CH_3OCH_3

4. Which compound is an isomer of C_4H_9OH?

   (1) $C_3H_7CH_3$ (2) $C_2H_5OC_2H_5$ (3) $C_2H_5COOC_2H_5$ (4) CH_3COOH

5. As the number of carbon atoms in a hydrocarbon molecule increases, the number of possible isomers generally

   (1) decreases   (2) increases   (3) remains the same

6. In the alkane series, each molecule contains

   (1) only one double bond   (2) two double bonds
   (3) one triple bond   (4) all single bonds

7. Which is the structural formula of methane?

(1)
$$H-\overset{\displaystyle H}{\underset{\displaystyle H}{C}}-\overset{\displaystyle H}{\underset{\displaystyle H}{C}}-H$$

(2)
$$\begin{array}{cc} H & \quad H \\ \diagdown & \diagup \\ C=C \\ \diagup & \diagdown \\ H & \quad H \end{array}$$

(3)
$$H-\overset{\displaystyle H}{\underset{\displaystyle H}{C}} - \overset{\displaystyle HH-\overset{\displaystyle H}{C}-HH}{\underset{\displaystyle H}{C}} - \overset{\displaystyle}{\underset{\displaystyle H}{C}}-H$$

(4)
$$H-\overset{\displaystyle H}{\underset{\displaystyle H}{C}}-H$$

8. Which structural formula represents a saturated hydrocarbon?

(1)
$$H-\overset{\displaystyle H}{\underset{\displaystyle H}{C}}-\overset{\displaystyle H}{\underset{\displaystyle H}{C}}-H$$

(2)
$$\begin{array}{cc} H & \quad H \\ \diagdown & \diagup \\ C=C \\ \diagup & \diagdown \\ H & \quad H \end{array}$$

(3)
$$H-\overset{\displaystyle H}{\underset{\displaystyle H}{C}}-\overset{\displaystyle H}{\underset{\displaystyle H}{C}}-Cl$$

(4)
$$H-\overset{\displaystyle H}{\underset{\displaystyle H}{C}}-\overset{\displaystyle O}{\overset{\|}{C}}-Cl$$

9. What is the formula of pentene?

(1) C_4H_8 (2) C_4H_{10} (3) C_5H_{10} (4) C_5H_{12}

10. What is the total number of pairs of electrons that one carbon atom shares with the other carbon atom in the molecule C_2H_4?

    (1) 1   (2) 2   (3) 3   (4) 4

11. Which formula represents an alkene?

    (1) CH_4 (2) C_2H_2 (3) C_3H_6 (4) C_4H_{10}

12. In which hydrocarbon series does each molecule contain one triple bond?

    (1) alkane   (2) alkene   (3) alkyne   (4) methane

13. Which set of formulas represents members of the same homologous series?

    (1) C, CH_4, CH_4O (2) C_2H_4, C_3H_6, C_4H_8
    (3) C_2H_2, C_2H_4, C_2H_6 (4) CH_2, CH_3, CH_4

14. Which hydrocarbon is a member of the series with the general formula C_nH_{2n-2}?

    (1) ethyne   (2) ethene   (3) butane   (4) pentene

15. Given the compound:

$$
\begin{array}{ccccccc}
 & H & H & H & H & \\
 & | & | & | & | & \\
H- & C & -C & =C & -C & -H \\
 & | & & & | & \\
 & H & & & H &
\end{array}
$$

    What is the general formula of the hydrocarbon series of which this compound is a member?
    (1) C_nH_{2n+2} (2) C_nH_{2n} (3) C_nH_{2n-2} (4) C_nH_{2n-6}

16. Given the compounds:

$$CH_3CH_2CH_2CH_3 \text{ and } CH_3CHCH_3$$
$$| $$
$$CH_3$$

These compounds are both

(1) alkynes
(3) isomers of butane

(2) alkenes
(4) isomers of propane

17. The compound $CH_3CH_2CH_2CH_3$ belongs to the series that has the general formula

(1) C_nH_{2n-2} (2) C_nH_{2n+2} (3) C_nH_{n-6} (4) C_nH_{n+6}

18. Which structural formula represents a saturated compound?

(1)
$$
\begin{array}{c}
H \\
 \\
 \\
H
\end{array}
C=C-C-H
$$

(2)   H−C≡C−C−H

(3)   H−C−C−C−H

(4)
C=C−C=C

19. Which kind of bond is most common in organic compounds?

(1) covalent   (2) ionic   (3) hydrogen   (4) electrovalent

OTHER ORGANIC COMPOUNDS

Other homologous series of organic compounds are formed by the replacement of one or more hydrogen atoms of a hydrocarbon by atoms of other elements. Members of these series are named from their corresponding hydrocarbons. However, they are not necessarily prepared directly from the compounds from which their names have been derived. Many of these groups have been classified according to the presence of some common particular arrangement of atoms known as functional groups. These groups provide characteristic properties to the compounds that contain them. See *Reference Tables for Physical Setting/Chemistry* Table R.

ALCOHOLS

The functional group for this class of organic compounds is the $(-OH)$ group. In alcohols, one or more hydrogens of a hydrocarbon have been replaced by this –OH group. Under ordinary conditions, no more than one –OH group can be attached to a single carbon atom. It should be noted that the –OH group of alcohols does not form the hydroxide ion in aqueous solutions. Therefore, alcohols are not bases.

Alcohols can be classified according to the number of ^-OH groups contained in each molecule. Monohydroxy (ℓ) alcohols contain one ^-OH group. Dihydroxy (ℓ) alcohols contain two ^-OH groups. Those alcohols containing three ^-OH groups are known as trihydroxy (ℓ) alcohols.

Since the functional group can be attached to any hydrocarbon, it is customary, when describing the classes of compounds to use the letter "R" to represent the rest of the molecule. Following this convention, the general formula for any alcohol is $R-O-H$.

In the IUPAC system, alcohols are named from the corresponding hydrocarbon by replacing the final -*e* with -*ol*. Methanol, CH_3OH, the simplest primary alcohol, is formed by replacing one hydrogen of methane with an –OH group. Ethanol, C_2H_5OH, is formed by replacing one hydrogen of ethane with an –OH group. The structural formulas of these two alcohols are

$$
\begin{array}{cc}
\begin{array}{c}
\text{H} \\
| \\
\text{H}-\text{C}-\text{OH} \\
| \\
\text{H}
\end{array}
&
\begin{array}{c}
\text{H}\quad\text{H} \\
|\quad\ | \\
\text{H}-\text{C}-\text{C}-\text{OH} \\
|\quad\ | \\
\text{H}\quad\text{H}
\end{array}
\\[2em]
\text{methanol} & \text{ethanol}
\end{array}
$$

Some other alcohols are propanol, C_3H_7OH, butanol, C_4H_9OH, and pentanol, $C_5H_{11}OH$. On longer chain alcohols, isomers are possible. The OH can be bonded to different carbons in the chain. As in the case with unsaturated hydrocarbons, the carbons in the chain are numbered from one end to the other. Start numbering from the end nearest the functional group. The number of the carbon with the OH preceeds the name of the alcohol.

$$\begin{array}{cccc} \text{OH} & \text{H} & \text{H} & \text{H} \\ | & | & | & | \\ \text{H}-\text{C}_1-\text{C}_2-\text{C}_3-\text{C}_4-\text{H} \\ | & | & | & | \\ \text{H} & \text{H} & \text{H} & \text{H} \end{array}$$

1-butanol

$$\begin{array}{cccc} \text{H} & \text{H} & \text{OH} & \text{H} \\ | & | & | & | \\ \text{H}-\text{C}_4-\text{C}_3-\text{C}_2-\text{C}_1-\text{H} \\ | & | & | & | \\ \text{H} & \text{H} & \text{H} & \text{H} \end{array}$$

2-butanol

ORGANIC ACIDS

The group of organic compounds known as the organic acids contain the functional group $-COOH$. Their general formula is RCOOH. Organic acids are named by replacing the final -e of the corresponding hydrocarbon with -oic, and adding the name acid. The first two members of this group are methanoic acid, HCOOH and ethanoic acid, C_2H_5COOH.

methanoic acid

ethanoic acid

ALDEHYDES

Aldehydes contain the functional group $-\overset{\overset{\textstyle H}{|}}{C}=O$ Their general formula is RCHO They are named by replacing the -e of the corresponding hydrocarbon with -al.

The functional group for aldehydes always comes at the end of the carbon chain. The first two aldehydes are methanal, HCHO, and ethanal, CH_3CHO.

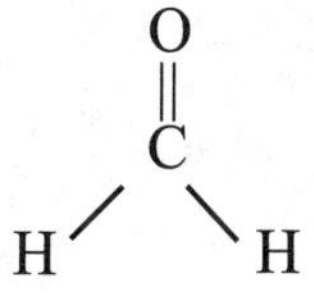

methanal

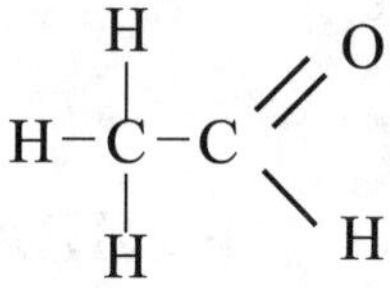

ethanal 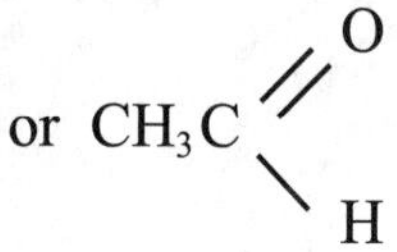   ethanal

KETONES

The general formula for ketones is

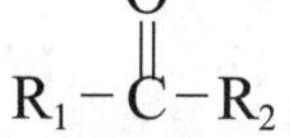

$$R_1 - \overset{\overset{\textstyle O}{\|}}{C} - R_2$$

The functional group is $-\overset{\overset{\textstyle O}{\|}}{C}-$ They are named by replacing the -e of the corresponding hydrocarbon with -one.

The simplest ketone is one in which both R_1 and R_2 are methyl (CH_3) groups.

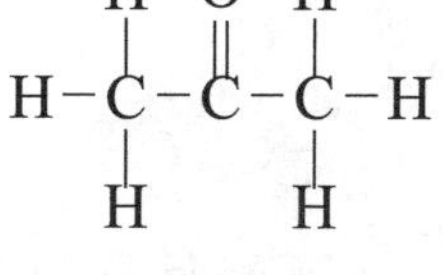

propanone

Ketones are isomers of aldehydes. The functional group is at the end of the chain for aldehydes but in the middle for ketones. Other examples of ketones are

$$H-\overset{\overset{\displaystyle H}{|}}{\underset{\underset{\displaystyle H}{|}}{C}}-\overset{\overset{\displaystyle O}{||}}{C}-\overset{\overset{\displaystyle H}{|}}{\underset{\underset{\displaystyle H}{|}}{C}}-\overset{\overset{\displaystyle H}{|}}{\underset{\underset{\displaystyle H}{|}}{C}}-\overset{\overset{\displaystyle H}{|}}{\underset{\underset{\displaystyle H}{|}}{C}}-H \quad \text{and} \quad H-\overset{\overset{\displaystyle H}{|}}{\underset{\underset{\displaystyle H}{|}}{C}}-\overset{\overset{\displaystyle H}{|}}{\underset{\underset{\displaystyle H}{|}}{C}}-\overset{\overset{\displaystyle O}{||}}{C}-\overset{\overset{\displaystyle H}{|}}{\underset{\underset{\displaystyle H}{|}}{C}}-\overset{\overset{\displaystyle H}{|}}{\underset{\underset{\displaystyle H}{|}}{C}}-H$$

2-pentanone   3-pentanone

ETHERS

The general formula for the ethers is $R_1 - O - R_2$. The functional group is —O—. Diethyl ether, $C_2H_5OC_2H_5$, is commonly used as a solvent and anesthetic.

$$H-\overset{\overset{\displaystyle H}{|}}{\underset{\underset{\displaystyle H}{|}}{C}}-\overset{\overset{\displaystyle H}{|}}{\underset{\underset{\displaystyle H}{|}}{C}}-O-\overset{\overset{\displaystyle H}{|}}{\underset{\underset{\displaystyle H}{|}}{C}}-\overset{\overset{\displaystyle H}{|}}{\underset{\underset{\displaystyle H}{|}}{C}}-H$$

diethyl ether

ESTERS

Esters have the general formula $R - \overset{\overset{\displaystyle O}{||}}{C} - O - R^1$. The Functional group is

$$-\overset{\overset{\displaystyle O}{||}}{C}-O-$$

similar to acids, but the oxygen is bonded to R^1 instead of hydrogen. The condensed structural formula would be $RCOOR^1$.

The ester's name uses both the R and the R^1 from its formula. The hydrocarbon R^1 comes first. The *–ane* ending is replaced by *–yl*. The acid resembling RCOO is named last. The *–ic* ending of the acid derivative is replaced by *–ate*.

In general: $RCOOR^1$ is $R^1–yl$ R *–ate*
          $C_2H_5COOCH_3$ is   methyl propanoate
          $HCOOC_2H_5$ is   ethyl methanoate

HALIDES

The halogens F, Cl, Br, and I are common addition or substitution products of hydrocarbons. There can be one, two (*di*), three (*tri*), four (*tetra*), or more halogens on one molecule. The halogen name shortens and ends in – o when used as a functional group. Some examples are:

chloromethane

1, 2 – diflouromethane

2-iodopropane

1,1,2-tribromobutane

AMINES

Amines have the functional group $-NH_2$. The general formula for an amines is RNH_2. They are named by dropping the $-e$ of the hydrocarbon followed by $-amine$. Some examples are:

methanamine   2-propanamine   1,2-butandiamine

AMINO ACIDS

Amino acids combine the functional group of acids and amines on the same molecule. They have the general formula :

$$R-\underset{H}{\overset{NH_2}{C}}-COOH.$$

Proteins are formed by reacting amino acid forming long chains. Simple amino acids can be named using the IUPAC system, but most are complex and common names are used.

2 – aminopropanoic acid

AMIDES

Amides combine the functional groups of aldehydes and amines on the same carbon. The general formula for amides is

$$
\begin{array}{c}
O \\
\parallel \\
R - C - NH_2
\end{array}
$$

The name has the ending -*amide* added to the hydrocarbon after the -*e* is dropped.

ethanamide

CHARACTERISTICS OF ORGANIC COMPOUNDS

Organic compounds have all covalent bonds (shared electron pairs). Their physical and chemical properties are generally those of molecular compounds, as contrasted to ionic compounds.

Some important properties include the following:

1.   Organic compounds are generally nonpolar.

2.   Only a few organic compounds will dissolve in water. These include ethanoic acid, various sugars, and certain alcohols. Acids, alcohols, and sugars have −OH on them as does water. They form hydrogen bonds to water which is a relatively strong intermolecular force. Many organic compounds are soluble in nonpolar solvents. These solvents are usually organic compounds themselves.

3.   Most organic compounds are nonelectrolytes. Organic acids are exceptions — they are weak electrolytes.

4.   Organic compounds have low melting points, due to the fact that the compounds are held together by weak intermolecular forces.

5.   Reaction rates of organic compounds are slower than those of inorganic compounds. In contrast to their weak intermolecular forces, the covalent bonds within the organic molecules are very strong. Activation energy, therefore, is very high, and catalysts are often used to increase reaction rates.

$$\boxed{\textbf{QUESTIONS}}$$

Answer the following questions using Tables P, Q, and R of the *Reference Tables for Physical Setting/Chemistry.*

1. Which compound is an electrolyte?

(1) C_2H_5OH (2) $C_3H_5(OH)_3$ (3) CH_3OH (4) CH_3COOH

2. Which is the formula of methanal?

(1)
```
      O
     //
 H−C
     \
      H
```

(2)
```
    H
    |
 H−C−OH
    |
    H
```

(3)
```
      O
     //
 H−C
     \
      OH
```

(4)
```
    H
    |
 H−C−H
    |
    H
```

3. Which formula represents an organic acid?

(1) $HCOOCH_3$ (2) CH_3CH_2OH (3) CH_3COCH_3 (4) $HCOOH$

4. Which compound is an isomer of propanone

$$\underset{\underset{\displaystyle H}{|}}{\overset{\overset{\displaystyle H}{|}}{H-C}} - \underset{\underset{\displaystyle O}{\|}}{C} - \underset{\underset{\displaystyle H}{|}}{\overset{\overset{\displaystyle H}{|}}{C}} - H?$$

(1) $H-\underset{\underset{\displaystyle H}{|}}{\overset{\overset{\displaystyle H}{|}}{C}}-\underset{\underset{\displaystyle H}{|}}{\overset{\overset{\displaystyle H}{|}}{C}}-C\underset{\displaystyle H}{\overset{\displaystyle O}{\diagup}}$

(2) $H-\underset{\underset{\displaystyle H}{|}}{\overset{\overset{\displaystyle H}{|}}{C}}-\underset{\underset{\displaystyle H}{|}}{\overset{\overset{\displaystyle H}{|}}{C}}-C\underset{\displaystyle OH}{\overset{\displaystyle O}{\diagup}}$

(3) $H-\underset{\underset{\displaystyle H}{|}}{\overset{\overset{\displaystyle H}{|}}{C}}-\underset{\underset{\displaystyle H}{|}}{\overset{\overset{\displaystyle H}{|}}{C}}-\underset{\underset{\displaystyle H}{|}}{\overset{\overset{\displaystyle H}{|}}{C}}-OH$

(4) $H-\underset{\underset{\displaystyle H}{|}}{\overset{\overset{\displaystyle H}{|}}{C}}-\underset{\overset{\displaystyle O}{\|}}{C}-H$

5. Which formula could represent 1,2-ethanediol?

(1) $C_2H_4(OH)_2$ (2) $C_3H_5(OH)_3$ (3) $Ca(OH)_2$ (4) $Co(OH)_3$

6. Which class of compounds has the general formula R_1-O-R_2?

(1) esters   (2) alcohols   (3) ethers   (4) aldehydes

7. What is the minimum number of carbon atoms a ketone may contain?

(1) 1   (2) 2   (3) 3   (4) 4

8. Which is the formula for methanoic acid?

(1) CH_3OH (2) C_2H_5OH (3) $HCOOH$ (4) $HC_2H_3O_2$

9. Which structural formula represents an aldehyde?

(1)
$$
\begin{array}{c}
\text{H} \quad \text{O} \\
| \quad\; || \\
\text{H}-\text{C}-\text{C}-\text{OH} \\
| \\
\text{H}
\end{array}
$$

(2)
$$
\begin{array}{c}
\text{H} \;\; \text{H} \;\; \text{H} \\
| \quad | \quad | \\
\text{H}-\text{C}-\text{C}-\text{C}-\text{H} \\
| \quad | \quad | \\
\text{H} \;\; \text{H} \;\; \text{H}
\end{array}
$$

(3)
$$
\begin{array}{c}
\text{H} \quad \text{O} \\
| \quad\; || \\
\text{H}-\text{C}-\text{C}-\text{H} \\
| \\
\text{H}
\end{array}
$$

(4)
$$
\begin{array}{c}
\text{H} \;\; \text{H} \\
| \quad | \\
\text{H}-\text{C}-\text{C}-\text{OH} \\
| \quad | \\
\text{H} \;\; \text{H}
\end{array}
$$

10. Molecules of 1-propanol and 2-propanol have different

(1) percentage compositions   (2) molecular masses
(3) molecular formulas   (4) structural formulas

11. In an aqueous solution, which compound will be acidic?

(1) CH_3COOH (2) CH_3CH_2OH (3) $C_3H_5(OH)_3$ (4) CH_3OH

12. Which structural formula represents a compound that is an isomer of

$$H-\underset{\underset{H}{|}}{\overset{\overset{H}{|}}{C}}-\underset{\underset{H}{|}}{\overset{\overset{H}{|}}{C}}-\underset{\underset{H}{|}}{\overset{\overset{H}{|}}{C}}-\underset{\underset{Br}{|}}{\overset{\overset{Br}{|}}{C}}-H \ ?$$

(1)
$$H-\underset{\underset{Br}{|}}{\overset{\overset{Br}{|}}{C}}-\underset{\underset{H}{|}}{\overset{\overset{H}{|}}{C}}-\underset{\underset{H}{|}}{\overset{\overset{H}{|}}{C}}-\underset{\underset{H}{|}}{\overset{\overset{H}{|}}{C}}-\underset{\underset{H}{|}}{\overset{\overset{H}{|}}{C}}-H$$

(2)
$$H-\underset{\underset{H}{|}}{\overset{\overset{Br}{|}}{C}}-\underset{\underset{H-C-H}{|}}{\overset{\overset{Br}{|}}{C}}-\underset{\underset{H}{|}}{\overset{\overset{Br}{|}}{C}}-H$$

(3)
$$H-\underset{\underset{H}{|}}{\overset{\overset{H}{|}}{C}}-\underset{\underset{H}{|}}{\overset{\overset{Br}{|}}{C}}-\underset{\underset{H}{|}}{\overset{\overset{H}{|}}{C}}-\underset{\underset{H}{|}}{\overset{\overset{H}{|}}{C}}-Br$$

(4)
$$Br-\underset{\underset{H-C-H}{|}}{\overset{\overset{H}{|}}{C}}-Br$$

13. A general characteristic of organic compounds is that they all

(1) react vigorously   (2) dissolve in water
(3) are strong electrolytes   (4) melt at relatively low temperatures

14. Compared with the rate of an inorganic reaction the rate of an organic reaction is usually

(1) faster, because the organic particles are ions
(2) faster, because the organic particles are molecules
(3) slower, because the organic particles are ionic
(4) slower because the organic particles are molecules

15. Which is an isomer of $CH_3CH_2CH_2COOH$?

(1) $CH_3CH_2OCH_2CH_3$ (2) $CH_3CH_2CH_2OCH_3$
(3) $CH_3CH_2CH_2CH_2OH$ (4) $CH_3COOCH_2CH_3$

16. Which of the following is and amino acid?

    (1) $C_2H_5CH(NH_2)COOH$

    (2) HCOOH

    (3) $CH_3-\overset{\overset{\textstyle O}{\|}}{C}-NH_2$

    (4) $CH_3CH_2NH_2$

17. What is the name of the following compound?

$$H-\overset{\overset{\textstyle H}{|}}{\underset{\underset{\textstyle H}{|}}{C}}-\overset{\overset{\textstyle H}{|}}{\underset{\underset{\textstyle Br}{|}}{C}}-\overset{\overset{\textstyle Br}{|}}{\underset{\underset{\textstyle H}{|}}{C}}-H$$

    (1) bromopropane
    (3) 3,4-dibromopropane

    (2) 1,2-dibromopropane
    (4) 3,4-dibromoethane

18. What is the class of the following compound?

$$H-\overset{\overset{\textstyle H}{|}}{N}-\overset{\overset{\textstyle O}{\|}}{C}-C_3H_7$$

    (1) amine   (2) acid   (3) amide   (4) a ldehyde

19. Which of the following is a 2-butanamine?

    (1) $H-\overset{\overset{\textstyle NH_2}{|}}{\underset{\underset{\textstyle H}{|}}{C}}-\overset{\overset{\textstyle NH_2}{|}}{\underset{\underset{\textstyle H}{|}}{C}}-\overset{\overset{\textstyle H}{|}}{\underset{\underset{\textstyle H}{|}}{C}}-\overset{\overset{\textstyle H}{|}}{\underset{\underset{\textstyle H}{|}}{C}}-H$

    (2) $H-\overset{\overset{\textstyle H}{|}}{\underset{\underset{\textstyle H}{|}}{C}}-\overset{\overset{\textstyle H}{|}}{\underset{\underset{\textstyle H}{|}}{C}}-\overset{\overset{\textstyle NH_2}{|}}{\underset{\underset{\textstyle H}{|}}{C}}-\overset{\overset{\textstyle H}{|}}{\underset{\underset{\textstyle H}{|}}{C}}-H$

    (3) $H-\overset{\overset{\textstyle H}{|}}{\underset{\underset{\textstyle H}{|}}{C}}-\overset{\overset{\textstyle H}{|}}{\underset{\underset{\textstyle H}{|}}{C}}-\overset{\overset{\textstyle H}{|}}{\underset{\underset{\textstyle H}{|}}{C}}-NH_2$

    (4) $H-\overset{\overset{\textstyle H}{|}}{\underset{\underset{\textstyle H}{|}}{C}}-\overset{\overset{\textstyle H}{|}}{\underset{\underset{\textstyle NH_2}{|}}{C}}-\overset{\overset{\textstyle H}{|}}{\underset{\underset{\textstyle H}{|}}{C}}-H$

ORGANIC REACTIONS

As noted earlier, organic reactions typically proceed at much slower rates than do inorganic reactions. For this reason, the use of catalysts is a common practice. In many organic reactions, only the functional group is involved. The greater part of the reacting molecules remain unchanged during the course of the reaction, and can easily be identified in the products.

Substitution

As the name implies, substitution reactions involve replacing one kind of atom or group with another kind of atom or group. For the saturated hydrocarbons, all substitution reactions (except for the special cases of combustion and thermal decomposition) involve replacement of hydrogen atoms. The halogen (F, Cl, Br, I) derivatives of the alkanes can be prepared by substitution reactions between the alkane and the halogen. The general term for these reactions is halogen *substitution*.

$$
\begin{array}{cccc}
\text{H}-\overset{\displaystyle H}{\underset{\displaystyle H}{\text{C}}}-\overset{\displaystyle H}{\underset{\displaystyle H}{\text{C}}}-\text{H} + \text{Br}_2 & \rightarrow & \text{H}-\overset{\displaystyle H}{\underset{\displaystyle H}{\text{C}}}-\overset{\displaystyle H}{\underset{\displaystyle H}{\text{C}}}-\text{Br} + \text{HBr}
\end{array}
$$

ethane	bromine	bromoethane	hydrogen bromide

Preparation of the halogen derivatives by substitution always results in a by-product of the hydrogen halide.

Addition

Addition reactions involve *adding* two or more atoms to carbon atoms that are attached to other carbon atoms by double or triple bonds. Thus, addition reactions are generally limited to the unsaturated hydrocarbons. Addition reactions take place more easily than substitution reactions. Their rates are often as fast as those of ionic reactions. As a result, unsaturated compounds are considered more reactive than saturated compounds. Furthermore, those with triple bonds (alkynes) tend to be more reactive than those with double bonds (alkenes). Addition of hydrogen to an unsaturated compound, however, usually requires the presence of a catalyst and an elevated temperature. The hydrogen addition reaction is called *hydrogenation.* Addition reactions between unsaturated hydrocarbons and chlorine and bromine to produce halogen derivatives take place at room temperature.

$$ \underset{\text{ethene}}{\overset{\displaystyle H}{\underset{\displaystyle H}{}}\!\!\!\!C=C\overset{\displaystyle H}{\underset{\displaystyle H}{}}} \quad + \quad \underset{\text{bromine}}{Br_2} \quad \rightarrow \quad \underset{\text{1,2–dibromoethane}}{H-\overset{\displaystyle H}{\underset{\displaystyle Br}{C}}-\overset{\displaystyle H}{\underset{\displaystyle Br}{C}}-H} $$

Addition reactions are characterized by the formation of a single product. This is in contrast with substitution reactions in which more than one product is typical.

Fermentation

Fermentation is a process ordinarily associated with living systems. Enzymes produced by the living organisms serve as catalysts for the reactions in which organic molecules are broken down.

For example, the fermentation of glucose is shown:

$$ \underset{\text{glucose}}{C_6H_{12}O_6} \quad \underset{\text{(from yeast)}}{\overset{\text{zymase}}{\rule{3cm}{0.4pt}}} \quad \underset{\text{ethanol}}{2C_2H_5OH} \quad + \quad \underset{\text{carbon dioxide}}{2CO_2} $$

Esterification

Esterification derives its name from the name of the products, esters. Esterification involves the reaction between an organic acid and an alcohol to produce an ester and water.

Esters have a first and last name. The first name is derived from the alcohol name with a *-yl* ending. The last name comes from the organic acid with an *-ate* ending.

$$ \underset{\text{methanol}}{H-\overset{\displaystyle H}{\underset{\displaystyle H}{C}}-O-H} \quad + \quad \underset{\text{ethanoic acid}}{H-O-\overset{\displaystyle O}{C}-\overset{\displaystyle H}{\underset{\displaystyle H}{C}}-H} \quad \rightarrow \quad \underset{\substack{\text{methyl ethanoate}\\\text{(ester)}}}{H-\overset{\displaystyle H}{\underset{\displaystyle H}{C}}-O-\overset{\displaystyle O}{C}-\overset{\displaystyle H}{\underset{\displaystyle H}{C}}-H} \quad + \quad H_2O $$

Since esterification involves an acid and an –OH group, it is often compared with neutralization of inorganic acids with bases, producing salts and water. (See p. 163) Esterification, however, is not an ionic reaction, and esters are covalent compounds. Esterification is a slow reaction, usually requiring a catalyst, and it is reversible. Esters are responsible for the aromas associated with many fruits, flowers, and leaves. Lipids (fats and oils) are esters formed from esterification of glycerol (1, 2, 3-propantriol) by long-chain organic acids (fatty acids).

Saponification

The hydrolysis of fats (complex esters) by bases is called *saponification*. The organic salts that are produced are soaps. Glycerol, an alcohol, is a second product of saponification reactions and is considered a byproduct in the manufacture of soap.

$$
\begin{array}{c}
H \quad\quad O \\
| \quad\quad\; \| \\
H-C-O-C-C_{17}H_{35} \\
| \quad\quad\; O \\
\quad\quad\quad \| \\
H-C-O-C-C_{17}H_{35} \quad + \quad 3NaOH \\
| \quad\quad\; O \\
\quad\quad\quad \| \\
H-C-O-C-C_{17}H_{35} \\
| \\
H
\end{array}
\quad\longrightarrow\quad
\begin{array}{c}
H \\
| \\
H-C-OH \\
| \\
H-C-OH \\
| \\
H-C-OH \\
| \\
H
\end{array}
\quad + \quad 3NaOCC_{17}H_{35}
$$

animal fat   glycerol   soap

"Like dissolves like" is a useful saying. Polar solutes dissolve best in polar solvents (like water), and non-polar solutes dissolve best in non-polar solvents (like hydrocarbons). Soap can bring non-polar solutes together with polar solvents because it is both. One end of the soap molecule is a long hydrocarbon chain attracting it to non-polar stains like grease.

The other end ($-\overset{\overset{\textstyle O}{\|}}{C}-ONa$) is polar and attracted to water.

Combustion

Saturated hydrocarbons react readily with oxygen under conditions of combustion. Such reactions result in the oxidation of the carbon to carbon monoxide or carbon dioxide, depending on the amount of oxygen available. The oxygen bonds with hydrogen from the hydrocarbon to form water as well as the oxide of carbon. Oxidation reactions have great significance because of the liberation of energy associated with them. Energy is derived from fuels by combustion and from food by cellular respiration, both processes involving oxidation reactions.

$$CH_4 + O_2 \rightarrow CO_2 + H_2O + \text{heat}$$

methane   carbon   water
                   dioxide

Polymerization

Polymerization is a name given to reactions in which large molecules are made from smaller molecules. Polymerization occurs in nature in the production of proteins and starches by living organisms. Synthetic rubbers, plastics, and fibers are results of polymerization reactions.

Polymers are composed of many repeating units, called monomers, which are joined together by one of two types of polymerization reactions—condensation or addition.

Condensation. Condensation polymerization results from joining monomers by dehydration. It is sometimes called dehydration synthesis (because of the removal of water).

$$HO-\overset{\displaystyle H}{\underset{\displaystyle H}{C}}-\overset{\displaystyle H}{\underset{\displaystyle H}{C}}-OH \ + \ HO-\overset{\displaystyle H}{\underset{\displaystyle H}{C}}-\overset{\displaystyle H}{\underset{\displaystyle H}{C}}-OH \ \rightarrow \ HO-\overset{\displaystyle H}{\underset{\displaystyle H}{C}}-\overset{\displaystyle H}{\underset{\displaystyle H}{C}}-O-\overset{\displaystyle H}{\underset{\displaystyle H}{C}}-\overset{\displaystyle H}{\underset{\displaystyle H}{C}}-OH \ + \ H_2O$$

monomer   monomer   dimer (polymer)

This process may be repeated to produce a long-chain polymer. Monomers involved in condensation must have at least two functional groups. Examples of condensation polymers include silicones, polyesters, polyamides, phenolic plastics, and nylons.

Addition. Addition polymerization, as do all addition reactions, involves opening of double and triple bonds of unsaturated hydrocarbons.

Vinyl plastics, such as polyethylene and polystyrene, are examples of addition polymers.

$$n \left(\begin{array}{c} H \quad H \\ | \quad | \\ H - C = C - H \\ \end{array} \right) \rightarrow \left(\begin{array}{c} H \quad H \\ | \quad | \\ - C - C - \\ | \quad | \\ H \quad H \\ \end{array} \right)_n$$

| * n stands for a large number |

| ethene | polyethylene or polyethene |
| (the monomer) | (the polymer) |

QUESTIONS

Answer the following questions using Tables P, Q, and R of the *Reference Tables for Physical Setting/Chemistry.*

1. The organic reaction

$$HCOOH + CH_3CH_2CH_2CH_2OH \rightarrow HCOOCH_2CH_2CH_2CH_3 + HOH,$$

is an example of

(1) fermentation
(3) polymerization

(2) esterification
(4) saponification

2. Alkanes differ from alkenes in that alkanes

(1) are hydrocarbons
(3) have the general formula C_nH_{2n}

(2) are saturated compounds
(4) undergo addition reactions

3. Which molecule is represented by X in the reaction

$$H - C \equiv C - H + 2Br_2 \rightarrow X?$$

(1)
$$\begin{array}{c} H \quad H \\ | \quad | \\ H - C - C - Br \\ | \quad | \\ H \quad H \end{array}$$

(2)
$$\begin{array}{c} H \quad H \\ | \quad | \\ H - C - C - H \\ | \quad | \\ Br \quad Br \end{array}$$

(3)
$$\begin{array}{c} H \quad H \\ | \quad | \\ H - C - C - Br \\ | \quad | \\ H \quad Br \end{array}$$

(4)
$$\begin{array}{c} Br \quad Br \\ | \quad | \\ H - C - C - H \\ | \quad | \\ Br \quad Br \end{array}$$

4. Which equation represents an esterification reaction?

(1) $C_6H_{12}O_6 \rightarrow 2C_2H_5OH + 2CO_2$
(2) $C_5H_{10} + H_2 \rightarrow C_5H_{12}$
(3) $C_3H_8 + Cl_2 \rightarrow C_3H_7Cl + HCl$
(4) $HCOOH + CH_3OH \rightarrow HCOOCH_3 + HOH$

5. What is the correct IUPAC name for

$$\begin{array}{ccc} H & Cl & Cl \\ | & | & | \\ H-C-&C-&C-H\,? \\ | & | & | \\ H & H & H \end{array}$$

(1) 1,2-dichlorobutane
(3) 1,2-dichloropropane
(2) 2,3-dichlorobutane
(4) 2,3-dichloropropane

6. Which equation represents a substitution reaction?

(1) $CH_4 + 2O_2 \rightarrow CO_2 + 2H_2O$
(2) $C_2H_4 + Br_2 \rightarrow C_2H_4Br_2$
(3) $C_3H_6 + H_2 \rightarrow C_3H_8$
(4) $C_4H_{10} + Cl_2 \rightarrow C_4H_9Cl + HCl$

7. Which reaction produces ethanol as one of the principal products?

(1) an esterification reaction
(3) a saponification reaction
(2) a neutralization reaction
(4) a fermentation reaction

8. Which is the product of the reaction between ethene and chlorine?

(1) $\begin{array}{cc} H & H \\ | & | \\ H-C-&C-Cl \\ | & | \\ H & H \end{array}$

(2) $\begin{array}{c} H \\ | \\ H-C-Cl \\ | \\ H \end{array}$

(3) $\begin{array}{cc} H & H \\ | & | \\ Cl-C-&C-Cl \\ | & | \\ H & H \end{array}$

(4) $\begin{array}{c} H \\ | \\ Cl-C-Cl \\ | \\ H \end{array}$

9. In which type of reaction are long-chain molecules formed from smaller molecules?

(1) substitution   (2) saponification
(3) fermentation   (4) polymerization

10. Cellulose is an example of

(1) a synthetic polymer   (2) a natural polymer
(3) an ester   (4) a ketone

11. A reaction between CH_3COOH and an alcohol produced the ester CH_3COOCH_3. The alcohol used in the reaction was

(1) CH_3OH (2) C_2H_5OH (3) C_3H_7OH (4) C_4H_9OH

12. An alcohol and an organic acid are combined to form water and a compound with a pleasant odor. This reaction is an example of

(1) saponification   (2) esterification   (3) polymerization   (4) fermentation

13. Given the equation:

$$H-O-\underset{\underset{H}{|}}{\overset{\overset{H}{|}}{C}}-\underset{\underset{H}{|}}{\overset{\overset{H}{|}}{C}}-O-H + H-O-\underset{\underset{H}{|}}{\overset{\overset{H}{|}}{C}}-\underset{\underset{H}{|}}{\overset{\overset{H}{|}}{C}}-O-H \rightarrow H-O-\underset{\underset{H}{|}}{\overset{\overset{H}{|}}{C}}-\underset{\underset{H}{|}}{\overset{\overset{H}{|}}{C}}-O-\underset{\underset{H}{|}}{\overset{\overset{H}{|}}{C}}-\underset{\underset{H}{|}}{\overset{\overset{H}{|}}{C}}-O-H + H_2O$$

Which type of reaction is represented?

(1) condensation polymerization   (2) addition polymerization
(3) esterification   (4) saponification

14. Which is the formula of 2,2-dichloropropane?

(1)
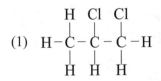

$$\begin{array}{c} \text{H} \quad \text{Cl} \quad \text{Cl} \\ | \quad\; | \quad\; | \\ \text{H}-\text{C}-\text{C}-\text{C}-\text{H} \\ | \quad\; | \quad\; | \\ \text{H} \quad \text{H} \quad \text{H} \end{array}$$

(2)
$$\begin{array}{c} \text{H} \quad \text{Cl} \quad \text{H} \\ | \quad\; | \quad\; | \\ \text{H}-\text{C}-\text{C}-\text{C}-\text{H} \\ | \quad\; | \quad\; | \\ \text{H} \quad \text{Cl} \quad \text{H} \end{array}$$

(3)
$$\begin{array}{c} \text{H} \quad \text{Cl} \quad \text{Cl} \quad \text{H} \\ | \quad\; | \quad\; | \quad\; | \\ \text{H}-\text{C}-\text{C}-\text{C}-\text{C}-\text{H} \\ | \quad\; | \quad\; | \quad\; | \\ \text{H} \quad \text{H} \quad \text{H} \quad \text{H} \end{array}$$

(4)
$$\begin{array}{c} \text{H} \quad \text{Cl} \quad \text{H} \quad \text{H} \\ | \quad\; | \quad\; | \quad\; | \\ \text{H}-\text{C}-\text{C}-\text{C}-\text{C}-\text{H} \\ | \quad\; | \quad\; | \quad\; | \\ \text{H} \quad \text{Cl} \quad \text{H} \quad \text{H} \end{array}$$

15. The type of reaction represented by the equation

$$C_2H_4 + H_2 \rightarrow C_2H_6 \quad \text{is called}$$

(1) substitution   (2) polymerization   (3) addition   (4) esterification

16. Which reaction is used to produce polyethylene $(C_2H_4)_n$ from ethylene?

(1) addition polymerization   (2) substitution
(3) condensation polymerization   (4) reduction

17. The process of opening double bonds and joining monomer molecules to form polyvinyl chloride is called

(1) addition polymerization   (2) condensation polymerization
(3) dehydration polymerization   (4) neutralization polymerization

18. Which hydrocarbon will undergo a substitution reaction with chlorine?

(1) methane   (2) ethyne   (3) propene   (4) butene

PRACTICE FOR CONSTRUCTED RESPONSE

1. Peanut oil is unsaturated. In making popular brands of peanut butter, the peanut oil is saturated (hydrogenated).

   a. What kind of chemical reaction changes unsaturated oil to saturated? Give an example. (2 points)

   _____

   b. Although saturated fats may be less healthy than unsaturated oils, why do some companies and their customers prefer saturated fats? (1 point)

   _____

   _____

   _____

2. Many artificial flavors are esters that are manufactured. They are the same chemical that occurs naturally in the fruit. One such ester is ethyl butanoate, the flavor of pineapple.

   a. Write an equation for the esterification of ethyl butanoate. (4 points)

   _____

   b. Name the reactants. (2 points)

   _____

3. Draw structural formulas and name *all* isomers of the following:

   a. C_2H_6O (4 points)

   _____

   b. C_2H_7N (4 points)

   _____

4. Draw structural formulas for the following compounds:
(4 points)

a. 2–pentene

b. butanoic acid

c. 3–hexanone

d. 2, 3–dichloroheptane

5. List all the errors in this structural formula:
(3 points)

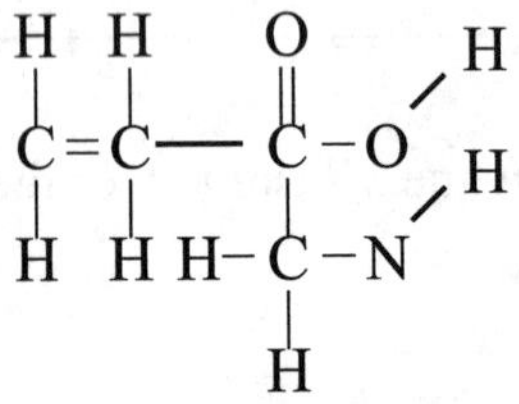

_____

_____

_____

6. Match the type of reaction to the correct reaction.
(7 points)

(1) addition   1. _____

(2) substitution   2. _____

(3) esterification   3. _____

(4) polymerization   4. _____

(5) fermentation   5. _____

(6) saponification   6. _____

(7) combustion   7. _____

a. $(C_{17}H_{35}COOCH)_3H_2 + 3Na\,OH \rightarrow 3C_{17}H_{35}COONa + H_2(CHOH)_3$

b. $CH_3OH + CH_3COOH \rightarrow CH_3COOCH_3 + H_2O$

c. $CH_4 + Cl_2 \rightarrow CH_3Cl + HCl$

d. $CH_4 + 2O_2 \rightarrow CO_2 + 2H_2O$

e. $C_2H_4 + Cl_2 \rightarrow CH_2Cl\,CH_2Cl$

f. $C_6H_{12}O_6 \xrightarrow{\text{zymase}} 2C_2H_5OH + 2CO_2$

g.

$$n\,H-\underset{\underset{H}{|}}{\overset{\overset{H}{|}}{N}}-\underset{\underset{R}{|}}{\overset{\overset{O}{\parallel}}{C}}-\overset{O}{\overset{\parallel}{C}}-OH + n\,H-\underset{\underset{H}{|}}{\overset{\overset{H}{|}}{N}}-\underset{\underset{R^1}{|}}{C}-\overset{O}{\overset{\parallel}{C}}-OH \rightarrow \left[-\underset{\underset{H}{|}}{\overset{\overset{H}{|}}{N}}-\underset{\underset{R}{|}}{C}-\overset{O}{\overset{\parallel}{C}}-\underset{\underset{H}{|}}{\overset{\overset{H}{|}}{N}}-\underset{\underset{R^1}{|}}{C}-\overset{O}{\overset{\parallel}{C}}- \right]_n$$

CHAPTER 8
OXIDATION - REDUCTION

REDOX

Redox is a term used for oxidation-reduction reactions. The single term redox is useful because both processes—oxidation and reduction—always occur together in the same reaction. Redox reactions result from the competition for electrons between atoms.

Oxidation

Oxidation is defined as the loss, or apparent loss, of electrons.

$$Zn(s) \rightarrow Zn^{2+}(aq) + 2e^-$$

The term is used in referring to any chemical change in which there is an increase in oxidation number (becomes more positive, or less negative). The particle that experiences the loss of electrons (and increase in oxidation number) is said to have been *oxidized*.

Reduction

Reduction is defined as the gain, or apparent gain, of electrons, with an accompanying decrease in oxidation number.

$$2H^+(aq) + 2e^- \rightarrow H_2(g)$$

The particle that experiences the gain of electrons (and decreased oxidation number) is said to have been *reduced* (becomes less positive, or more negative). The terms oxidation and reduction can be seen in the following reaction:

$$Zn(s) + 2H^+(aq) \rightarrow Zn^{2+}(aq) + H_2(g)$$

Zn loses electrons and is oxidized to Zn^{2+}

H+ gains electrons and is reduced to H_2.

Oxidation Number (Oxidation State)

Oxidation numbers are assigned to atoms or ions as convenient devices to keep track of electron transfers. The assignment is based on the arbitrary assumption that electrons shared between two atoms belong to the atom with the higher electronegativity. The oxidation number can be defined as a fictitious charge assigned to an atom or ion on the basis of a certain set of rules.

Rules for Assigning Oxidation Numbers

1. The oxidation number of each atom in free elements is always zero. The hydrogen in H_2, the sodium in Na, and the sulfur in S_8 all have oxidation numbers of zero.

2. The oxidation numbers of ions are the same as the charge on the ion. In $MgCl_2$ the Mg^{2+} ion has an oxidation number of +2. Each of the two Cl^- ions has an oxidation number of −1. In $FeCl_2$ the Fe^{2+} ion has an oxidation number of +2, while in $FeCl_3$ the Fe^{3+} ion has an oxidation number of +3.

> **Special Note:** It should be noted that when writing symbols for the charge on an ion, the numeral is written *before* the + or − sign (2+, 3+, 2−, etc.). When writing oxidation numbers, the + or − sign precedes the numeral (+1, +3, −1, etc.).

3. Since the Group 1 metals form only 1 + ions, their oxidation number is + 1 in all their compounds.

4. The metals in Group 2 form only 2+ ions, and their oxidation number is +2 in all their compounds.

5. Oxygen has a −2 oxidation number in practically all of its compounds. This rule is especially useful when identifying the oxidation numbers of elements in polyatomic ions. The few exceptions for oxygen are in peroxides, such as H_2O_2 and Na_2O_2, where the oxidation number is − 1 and in compounds with fluorine (OF_2), in which it is +2.

6. Hydrogen has an oxidation number of + 1 in all its compounds *except* metal hydrides formed with Group 1 and Group 2 metals (such as LiH and CaH_2 in which it is − 1).

7. The sum of the oxidation numbers in a compound must be zero.

8. The sum of the oxidation numbers in a polyatomic ion must be equal to the charge on the ion. In the CO_3^{2-} ion the three oxygens provide a total of −6. Since there is one carbon, it must contribute +4 in order for CO_3^{2-} to have a net oxidation number of −2.

Finding Oxidation Numbers

The following will illustrate practical methods of finding oxidation numbers.

1. Find the oxidation numbers of N and O in the compound N_2O. Using rule 2, oxygen has an oxidation number of − 2. Using rule 7, the total oxidation number of the nitrogen is +2. Thus, each nitrogen atom has an oxidation number of +1.

2. For compounds composed of more than two elements, such as KMnO$_4$:
   a. Construct a table similar to the one shown here.
   b. Use rule 3 for K and rule 5 for O. Write the oxidation numbers for these elements in the table.
   c. Multiply the oxidation numbers by the subscripts. Write the products in the bottom row of the table.
   d. Apply rule 7. The total oxidation number of Mn is +7. Since there is only 1 atom of Mn, the oxidation number for Mn in this compound is +7.

K	Mn	O$_4$
+1	?	-2
+1	+7	-8

QUESTIONS

Answer the following questions using the Periodic Table of the Elements in the *Reference Tables for Physical Setting/Chemistry.*

1. Oxidation-reduction reactions occur because of the competition between particles for

   (1) neutrons   (2) electrons   (3) protons   (4) positrons

2. All redox reactions involve

   (1) the gain of electrons, only
   (2) the loss of electrons, only
   (3) both the gain and the loss of electrons
   (4) neither the gain nor the loss of electrons

3. Which statement describes what occurs in the following redox reaction?

$$Cu(s) + 2Ag^+ (aq) \rightarrow Cu^{2+} (aq) + 2Ag (s)$$

   (1) Only mass is conserved.
   (2) Only charge is conserved.
   (3) Both mass and charge are conserved.
   (4) Neither mass nor charge is conserved.

4. In which compound does chlorine have the highest oxidation number?

   (1) KClO   (2) KClO$_2$ (3) KClO$_3$ (4) KClO$_4$

5. In which compound does sulfur have an oxidation number of −2?

   (1) SO_2 (2) SO_3 (3) Na_2S (4) Na_2SO_4

6. In which compound does chlorine have an oxidation number of +7?

   (1) $HClO_4$ (2) $HClO_3$ (3) $HClO_2$ (4) $HClO$

7. In the reaction $2CrO_4^{2-}$ (aq) + $2H^+$ (aq) → $Cr_2O_7^{2-}$ (aq) + $H_2O(\ell)$, the oxidation number of chromium

   (1) decreases   (2) increases   (3) remains the same

8. Which is a redox reaction?

   (1) $Mg + 2HCl \rightarrow MgCl_2 + H_2$
   (2) $Mg(OH)_2 + 2HCl \rightarrow MgCl_2 + 2H_2O$
   (3) Mg^{2+} (aq) + $2OH^-$ (aq) → $Mg(OH)_2$
   (4) $MgCl_2 + 6H_2O \rightarrow MgCl_2 \cdot 6H_2O$

9. In the reaction $Cl_2 + H_2O \rightarrow HClO + HCl$, the hydrogen is

   (1) oxidized, only   (2) reduced, only
   (3) both oxidized and reduced   (4) neither oxidized nor reduced

10. In the reaction $Zn + Fe^{2+} \rightarrow Zn^{2+} + Fe$, what is oxidized?

    (1) Zn (2) Fe^{2+} (3) Zn^{2+} (4) Fe

11. Which change in oxidation number represents reduction?

    (1) −1 to +1   (2) −1 to −2   (3) −1 to +2   (4) −1 to 0

12. As an S^{2-} ion is oxidized to an S^0 atom, the number of protons in its nucleus

    (1) decreases   (2) increases   (3) remains the same

13. In the reaction $4NH_3 + 5O_2 \rightarrow 4NO + 6H_2O$, the oxidation number of nitrogen changes from

    (1) −2 to−3   (2) −2 to +3   (3) −3 to−2   (4) −3 to +2

14. Which is an oxidation-reduction reaction?

(1) $4Na + O_2 \rightarrow 2Na_2O$ (2) $3O_2 \rightarrow 2O_3$

(3) $AgNO_3 + NaCl \rightarrow AgCl + NaNO_3$ (4) $KI \rightarrow K^+ + I^-$

15. If element X forms the oxides XO and X_2O_3, the oxidation numbers of element X are

(1) +1 and +2   (2) +2 and +3   (3) +1 and +3   (4) +2 and +4

16. Given the oxidation-reduction reaction:

$$H_2 + 2Fe^{3+} \rightarrow 2H^+ + 2Fe^{2+}$$

Which species undergoes reduction?

(1) H_2 (2) Fe^{3+} (3) H^+ (4) Fe^{2+}

17. In the reaction $4Zn + 10HNO_3 \rightarrow 4Zn(NO_3)_2 + NH_4NO_3 + 3H_2O$, the zinc is

(1) reduced and the oxidation number changes from 0 to +2
(2) oxidized and the oxidation number changes from 0 to +2
(3) reduced and the oxidation number changes from +2 to 0
(4) oxidized and the oxidation number changes from +2 to 0

18. Which is a redox reaction?

(1) $2KBr + F_2 \rightarrow 2KF + Br_2$
(2) $2HCl + Mg(OH)_2 \rightarrow 2HOH + MgCl_2$
(3) $2NaCl + H_2SO_4 \rightarrow Na_2SO_4 + 2HCl$
(4) $Ca(OH)_2 + Pb(NO_3)_2 \rightarrow Ca(NO_3)_2 + Pb(OH)_2$

19. In the reaction $Zn(s) + Cu^{2+}(aq) \rightarrow Zn^{2+}(aq) + Cu(s)$, what is reduced?

(1) $Zn(s)$ (2) $Cu(s)$ (3) $Cu^{2+}(aq)$ (4) $Zn^{2+}(aq)$

20. In the reaction $2Al + 3Ni(NO_3)_2 \rightarrow 2Al(NO_3)_3 + 3Ni$, the aluminum is

(1) reduced and its oxidation number increases
(2) reduced and its oxidation number decreases
(3) oxidized and its oxidation number increases
(4) oxidized and its oxidation number decreases

21. Given the redox reaction: $Ni + Sn^{4+} \rightarrow Ni^{2+} + Sn^{2+}$
Which species has been oxidized?

(1) Ni (2) Sn^{4+} (3) Ni^{2+} (4) Sn^{2+}

22. In which substance is the oxidation number of nitrogen zero?

(1) N_2 (2) NH_3 (3) NO_2 (4) N_2O

23. What is the oxidation number of Pt in K_2PtCl_6?

(1) -2 (2) $+2$ (3) -4 (4) $+4$

24. In the reaction $2H_2S + 3O_2 \rightarrow 2SO_2 + 2H_2O$, what is reduced?

(1) oxygen   (2) water   (3) sulfur dioxide   (4) hydrogen sulfide

25. In the reaction $2Na + 2H_2O \rightarrow 2Na^+ + 2OH^- + H_2$, the substance oxidized is

(1) H_2 (2) H^+ (3) Na (4) Na^+

26. Given the reaction: $3Cu + 8HNO_3 \rightarrow 3Cu(NO_3)_2 + 2NO + 4H_2O$ what is oxidized?

(1) Cu^0 (2) N^{+5} (3) Cu^{+2} (4) N^{+2}

27. Given the reaction: $Sn^{2+}(aq) + 2Fe^{3+}(aq) \rightarrow Sn^{4+}(aq) + 2Fe^{2+}(aq)$ what is oxidized?

(1) Sn^{2+} (2) Fe^{3+} (3) Sn^{4+} (4) Fe^{2+}

28. The number of electrons transferred in the following reaction $Fe^{3+} \rightarrow Fe^{2+}$ is

(1) one, and iron is reduced   (2) two, and iron is reduced
(3) one, and iron is oxidized   (3) two, and iron is oxidized

29. Which equation is correctly written?

(1) $Pb^{2+} \rightarrow Pb + 2e^-$ (2) $Fe^{2+} + 2e^- \rightarrow Fe^{3+}$
(3) $Br_2 \rightarrow 2Br^- + 2e^-$ (4) $I_2 + 2e^- \rightarrow 2I^-$

30. What could X be in the equation $Cu^+ \rightarrow X + e^-$?

(1) Cu (2) Cu^{2+} (3) Cu^- (4) e^-

ELECTROCHEMISTRY

Half-reactions

All redox reactions consist of two "parts"—a gain of electrons (reduction) and a loss of electrons (oxidation). Each of these "parts" of a redox reaction can be thought of as a half-reaction. A separate equation, including the gain or loss of electrons, can be written for each half-reaction.

For example, consider the reaction:

$$Mg + Cl_2 \rightarrow MgCl_2$$

Written as half reactions, this reaction would be:

$$oxidation: \ Mg° \rightarrow Mg^{2+} + 2e^-$$
$$reduction: \ Cl_2 + 2e^- \rightarrow 2Cl^-$$

Half-cells

A **half-cell** is produced when a metal is placed in a solution of a salt of the metal. This arrangement provides a source of metal ions to be reduced and a source of metal atoms to be oxidized.

ELECTROCHEMICAL CELLS

There are two kinds of electrochemical cells. *Voltaic Cells* spontaneously change chemical energy into electrical energy. They are batteries. *Electrolytic cells* use electrical energy to produce chemical changes. This process is called electrolysis.

Voltaic Cells

Half-cells of two different metals can be used to produce electricity, provided the metals are connected by a wire to carry the electrons and the solutions are brought into contact by a salt bridge to carry the ions without mixing the solutions. Such a system is called a **voltaic cell**.

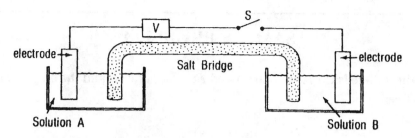

In voltaic cells, a spontaneous chemical reaction produces electrical energy.
Figure 8-1 Voltaic Cell

The redox chemical reaction that produces electricity is driven by the difference in *attraction for electrons* of the two metals in the voltaic cell. The metals are called **electrodes**. Electricity is moving electric charge. In wires, the charge moves as electrons. In solution, the charge moves as ions.

The relative attraction for electrons can be determined several ways. The higher the electronegativity and ionization energies (see Chapter 2) the more strongly the element attracts electrons. In redox reactions the **Activity Series** (Table J of the *References Tables for Physical Setting/Chemistry*) is generally used. The metals on Table J are listed in increasing order of attraction for electrons as one reads down the table. Those metals that are at the top are therefore most easily oxidized; Those metal ions that are at the bottom are most easily reduced. **For example:**

In figure 8-1   Let the left electrode be $Mg(s)$ and solution A be $Mg(NO_3)_2$ (aq). (Called a Mg/Mg^{2+} half-cell).
Let the right electrode $Cu(s)$ and solution B be $Cu(NO_3)_2$(aq) (a Cu/cu^{2+} half cell).
Since $Mg(s)$ is higher than $Cu(s)$ on Table J, $Mg(s)$ will be oxidized and Cu^{2+}(aq) will be reduced.

The half reactions are: $Mg(s) \rightarrow Mg^{2+}$ (aq) $+ 2e-$ oxidation half-reaction
Cu^{2+}(aq) $+ 2e- \rightarrow Cu(s)$ reduction half-reaction

The **direction of electric current** (electron flow) in the wire is toward the electrode which attracts electrons more (the lower one on the table). This electrode's ions will gain the electrons transferred and be reduced.

To keep the solutions of the two half-reactions neutral, ions must be able to flow from one to the other. This completes the electrical circuit. Completion can be achieved with a **salt bridge**, which is a tube containing a salt solution.

Electrodes

Cathode. The name *cathode* always applies to the electrode at which reduction occurs. In voltaic cells this is the positive electrode. In an electrolytic cell the cathode is the negative electrode.

Anode. The name *anode* always applies to the electrode at which oxidation occurs. In an electrolytic cell the anode is the positive electrode and in an voltaic cell, the anode is the negative electrode.

SAMPLE PROBLEM

Given the half-cells Cu°/Cu²⁺ (1.0M) and Zn°/Zn²⁺ (1.0M) and the diagram of the two half cells

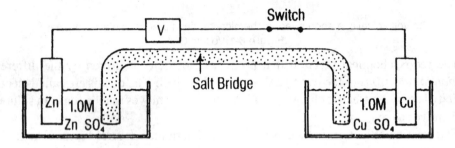

Figure 8-2

complete the following:

1. oxidation half-reaction =
   reduction half-reaction =

2. Label the anode and cathode

3. Show direction of electron flow

4. Show direction of ion flow

Solution

1. The table shows that the copper ion is more easily reduced. Therefore, the copper is the cathode, the zinc is the anode.
2. oxidation: $Zn° \rightarrow Zn^{2+} + 2e^-$
   reduction: $Cu^{2+} + 2e^- \rightarrow Cu°$
3. Electrons will flow from the anode (Zn) through the wire to the cathode (Cu).
4. In the salt bridge, positive ions flow toward the cathode and negative ions flow toward the anode.

Electrolytic Cells

In an **electrolytic cell**, electricity is used to produce a nonspontaneous chemical reaction. The reaction is called electrolysis. The energy required to proceed is usually supplied by an externally applied electric current (a battery).

The diagram below shows the electrolysis of fused (melted) sodium chloride.

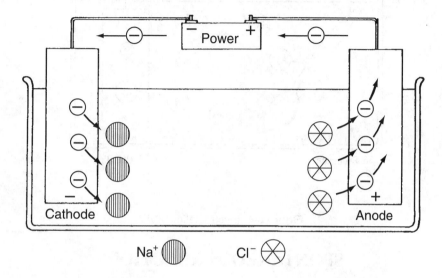

Figure 8-3. Na^+ ions are reduced, Cl^- ions are oxidized.

The two half-reactions are:

$$Na^+(aq) + e^- \rightarrow Na(s) \qquad 2Cl^- \rightarrow Cl_2(g) + 2e^-$$

The overall molecular reaction occurring in this cell can be represented as follows:

$$2NaCl \rightarrow 2Na(s) + Cl_2(g)$$

The electrolysis of fused sodium chloride is the only process for the production of sodium metal.

Electroplating is an example of electrolysis. In electrolysis one metal is plated onto another. In this process, an electric current is used to produce a chemical reaction. The material to be plated is the cathode; the metal used for plating is the anode. The electrolyte used is a salt of the anode. The diagram below shows the electroplating of a tin spoon with silver.

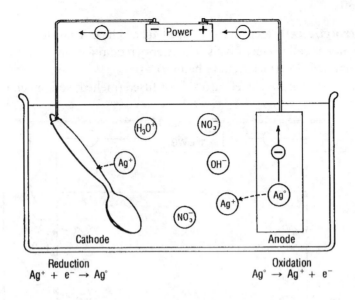

Figure 8-4. Electroplating silver.

SPONTANEOUS REACTIONS

The use of Reference Table J predicts which element is oxidized and which is reduced in a voltaic cell. If the reaction proceeds to completion once started, it is called spontaneous. Reference Table J can be used in the same way to make predictions of spontaneous reactions regardless of whether the reactions are separated into half-reactions or all reactants are mixed in the same place.

To predict spontaneous reactions for **metals** two criteria must be met:

1. You must have something to oxidize and something to reduce.
   Mg(s) and Al(s) can both only be oxidized.
   Mg^{2+}(aq) and Al^{3+}(aq) can both only be reduced.
   One metal must be neutral and the other and ion.

2. The metal being oxidized (the neutral one) must be higher on Table J.

   $Cu(s) + Ag(s) \rightarrow$ no reaction because neither can be reduced.
   $2Cr^{3+} + 3\,Mn(s) \rightarrow 3Mn^{2+} + 2Cr(s)$
   $Cr^{3+} + Ni(s) \rightarrow$ no reaction because Ni(s) is not higher.

To predict spontaneous reaction for **non-metals** using Table J, the criteria are similar to that for metals. The difference is that neutral non-metals can only gain electrons and be reduced. (Neutral metals lose electrons and are oxidized). Table J for non-metals shows F_2 attracting electrons the most and the most easily reduced. I_2 is the least easily reduced, but its ion I^- the most easily oxidized.

The criteria for spontaneous reaction among non-metals are:

1. You must have something to oxidize and something to reduce.

F_2 and Cl_2 can only be reduced.
F^- and Cl^- can only be oxidized.

2. The one being reduced (as written neutral on Table J) must be higher on the Table for the ion to be oxidized.

Cl_2 and Br_2 $\rightarrow$ no reaction because both can only be reduced
$Cl_2 + 2Br^- \rightarrow 2Cl^- + Br_2$
$I_2 + Br^- \rightarrow$ no reaction because I_2 is not higher on the Table.

In the reaction $3Ag^+(aq) + Cr(s) \rightarrow Cr^{3+}(aq) + 3Ag(s)$ The half-reactions are:

$Cr(s)^+ \rightarrow Cr^{3+} + 3e^-$ oxidation
$Ag^+(aq) + e^- \rightarrow Ag(s)$ reduction

We predict the reaction is spontaneous because the oxidation metal $Cr(s)$ is higher on Table J.

In the single replacement reaction

$$2Na(s) + Zn\,Cl_2(aq) \rightarrow Zn(s) + 2NaCl(aq)$$

Since chloride is always 1^-, the Zn ion is Zn^{2+} and the Na ion is Na^+ in their chlorides.

The oxidation half-reaction is $Na(s) \rightarrow Na^+(aq) + e^-$, and the reduction half-reaction is $Zn^{2+} + 2e^- \rightarrow Zn(s)$.

We predict the reaction will be spontaneous because the oxidized metal Na is higher on table J than Zn.

QUESTIONS

Answer the following questions using Table J of the *Reference Tables for Physical Setting/Chemistry.*

1. Given the reaction:

$$Ca(s) + Cu^{2+}(aq) \rightarrow Ca^{2+}(aq) + Cu(s)$$

What is the correct reduction half-reaction?

(1) $Cu^{2+}(aq) + 2e^- \rightarrow Cu(s)$ (2) $Cu^{2+}(aq) \rightarrow Cu(s) + 2e^-$
(3) $Cu(s) + 2e^- \rightarrow Cu^{2+}(aq)$ (4) $Cu(s) \rightarrow Cu^{2+}(aq) + 2e^-$

2. In the reaction $Mg + Cl_2 \rightarrow MgCl_2$, the correct half-reaction for the oxidation that occurs is

(1) $Mg + 2e^- \rightarrow Mg^{2+}$ (2) $Cl_2 + 2e^- \rightarrow 2Cl^-$
(3) $Mg \rightarrow Mg^{2+} + 2e^-$ (4) $Cl_2 \rightarrow 2Cl^- + 2e^-$

3. Based on Reference Table J, which of the following ions is most easily oxidized?

(1) F^- (2) Cl^- (3) Br^- (4) I^-

4. According to Reference Table J, which will reduce Mg^{2+} to $Mg(s)$?

(1) $Fe(s)$ (2) $Ba(s)$ (3) $Pb(s)$ (4) $Ag(s)$

5. Which ion will oxidize Fe?

(1) Zn^{2+} (2) Ca^{2+} (3) Mg^{2+} (4) Cu^{2+}

6. Which reaction will take place spontaneously?

(1) $Cu + 2H^+ \rightarrow Cu^{2+} + H_2$ (2) $2Au + 6H^+ \rightarrow 2Au^{3+} + 3H_2$
(3) $Pb + 2H^+ \rightarrow Pb^{2+} + H_2$ (4) $2Ag + 2H^+ \rightarrow 2Ag^+ + H_2$

Base your answers to questions 7 on the following reaction:

$$Mg(s) + 2Ag^+ (aq) \rightarrow Mg^{2+} (aq) + 2Ag(s)$$

7. Which species undergoes a loss of electrons?

(1) $Mg(s)$ (2) $Ag^+ (aq)$ (3) $Mg^{2+} (aq)$ (4) $Ag(s)$

Base your answers to questions 8 and 9 on the diagram of the chemical cell shown below. The reaction occurs at 1 atmosphere and 298 K.

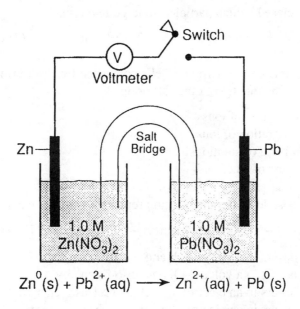

$$Zn^0(s) + Pb^{2+}(aq) \longrightarrow Zn^{2+}(aq) + Pb^0(s)$$

8. When the switch is closed, what occurs?

(1) Pb is oxidized and electrons flow to the Zn electrode.
(2) Pb is reduced and electrons flow to the Zn electrode.
(3) Zn is oxidized and electrons flow to the Pb electrode.
(4) Zn is reduced and electrons flow to the Pb electrode.

9. Which is the anode?

(1) $Pb^{2+}(aq)$ (2) $Pb(s)$ (3) $Zn^{2+}(aq)$ (4) $Zn(s)$

10. The diagram below represents an electrochemical cell.

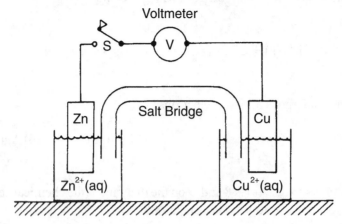

When switch S is closed, which particles undergo reduction?

(1) Zn^{2+} ions   (2) Zn atoms   (3) Cu^{2+} ions   (4) Cu atom

11. A chemical cell is made up of two half-cells connected by a salt bridge and an external conductor. What is the function of the salt bridge?

(1) to permit the migration of ions
(2) to prevent the migration of ions
(3) to permit the mixing of solutions
(4) to prevent the flow of electrons

12. Which statement best describes the reaction represented by the equation below?

$$2NaCl + 2H_2O + electricity \rightarrow Cl_2 + H_2 + 2NaOH$$

(1) The reaction occurs in a voltaic cell and releases energy.
(2) The reaction occurs in a voltaic cell and absorbs energy.
(3) The reaction occurs in an electrolytic cell and releases energy.
(4) The reaction occurs in an electrolytic cell and absorbs energy.

13. During the electrolysis of fused KBr, which reaction occurs at the positive electrode?

(1) Br^- ions are oxidized.   (2) Br^- ions are reduced.
(3) K^+ ions are reduced.   (4) K^+ ions are oxidized.

14. In an electrolytic cell, a Cl^- ion would be attracted to the

(1) positive electrode and oxidized
(2) positive electrode and reduced
(3) negative electrode and oxidized
(4) negative electrode and reduced

15. What occurs when an electrolytic cell is used for silverplating a spoon?

(1) A chemical reaction produces an electrical current.
(2) An electric current produces a chemical reaction.
(3) An oxidation reaction takes place at the cathode.
(4) A reduction reaction takes place at the anode.

16. The type of reaction in a voltaic cell is best described as a

(1) spontaneous oxidation reaction, only
(2) nonspontaneous oxidation reaction, only
(3) spontaneous oxidation-reduction reaction
(4) nonspontaneous oxidation-reduction reaction

Base your answers to questions 17 and 18 on the equation and diagram below which represents an electrochemical cell at 298 K and 1 atmosphere.

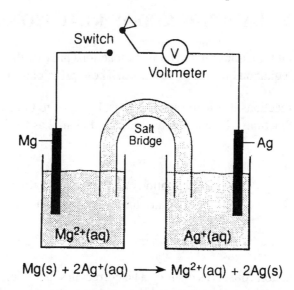

$$Mg(s) + 2Ag^+(aq) \longrightarrow Mg^{2+}(aq) + 2Ag(s)$$

17. Which species is oxidized when the switch is closed?

(1) $Mg(s)$ (2) Mg^{2+} (aq)   (3) $Ag(s)$ (4) $Ag^+(aq)$

18. When the switch is closed, electrons flow from

(1) $Mg(s)$ to $Ag(s)$ (2) Ag (s) to $Mg(s)$
(3) Mg^{2+} (aq) to Ag^+ (aq)   (4) Ag^+ (aq) to Mg^{2+} (aq)

19. In both the voltaic cell and the electrolytic cell, the anode is the electrode at which

(1) reduction occurs and electrons are lost
(2) reduction occurs and protons are lost
(3) oxidation occurs and electrons are lost
(4) oxidation occurs and protons are lost

20. Based on Reference Table J, which reaction will take place spontaneously?

(1) $Mg(s) + Ca^{2+} (aq) \rightarrow Mg^{2+} (aq) + Ca(s)$
(2) $Ba(s) + 2Na^+ (aq) \rightarrow Ba^{2+} (aq) + 2Na(s)$
(3) $Cl_2(g) + 2F^- (aq) \rightarrow 2Cl^- (aq) + F_2(g)$
(4) $I_2(g) + 2Br^-(aq) \rightarrow 2I^- (aq) + Br_2(g)$

BALANCING REDOX REACTIONS

Two fundamental principles related to redox reactions provide the basis for a structure to be used in balancing equations for redox reactions. These principles are:

1. In all redox reactions, the electrons lost must be equal to the electrons gained.
2. In all redox reactions, there is a conservation of charge as well as a conservation of mass.

These principles apply to redox equations and to half-reactions. For Example, in the reaction $Fe^{2+}(aq) + Al(s) \rightarrow Fe(s) + Al^{3+}(aq)$ the two half-reactions are:

$$Fe^{2+}(aq) \rightarrow Fe(s) \quad \text{and} \quad Al(s) \rightarrow Al^{3+}(aq)$$

Neither half-reaction conserves charge. The correct number of electrons must be added to one side of each equation to make the net charge add up to the same on both sides of the half reactions.

$$Fe^{2+}(aq) + 2e^- \rightarrow Fe(s)$$
$$Al(s) \rightarrow Al^{3+}(aq) + 3e^-$$

Both half-reactions show no net charge for the reactants and the products, so charge is conserved.

To balance the net equation, the electrons gained by the reduction half must equal the electrons lost by the oxidation half (multiply each half-reaction by the other half-reaction's number of electrons). Then add the two half-reactions together.

$$3(Fe^{2+} + 2e^- \rightarrow Fe) = 3Fe^{2+} + 6e^- \rightarrow 3Fe$$
$$2(Al \rightarrow Al^{3+} + 3e^-) = \underline{2Al \rightarrow 2Al^{3+} + 6e^-}$$
$$3Fe^{2+} + 2Al \rightarrow 2Al^{3+} + 3Fe$$

QUESTIONS

1. Given the reaction:

$$\_\_Hg^{2+} + \_\_Ag \rightarrow \_\_Hg + \_\_Ag^+$$

When the equation is completely balanced using the smallest whole-number coefficients, the coefficient of Hg will be

(1) 1   (2) 2   (3) 3   (4) 4

2. Given the unbalanced equation which represents aluminum metal reacting with an acid:

$$Al + H^+ \rightarrow Al^{+3} + H_2$$

What is the total number of moles of electrons lost by 1 mole of aluminum?

(1) 6   (2) 2   (3) 3   (4) 13

3. Given the unbalanced equation:

$$\_\_Fe + \_\_Ag^+ \rightarrow \_\_Ag + \_\_Fe^{3+}$$

When the equation is correctly balanced using smallest whole numbers, the coefficient of Ag+ is

(1) 5   (2) 2   (3) 3   (4) 4

4. Which of the following is a properly written oxidation half-reaction?

(1) $Mg - 2e^- \rightarrow Mg^{2+}$ (2) $Mg \rightarrow Mg^{2+} + 2e^-$

(3) $Mg^{2+} + 2e^- \rightarrow Mg$ (4) $Mg + 2e^- \rightarrow Mg^{2+}$

5. How many moles of electrons are needed to reduce one mole of Cu^{2+} to Cu^{1+}?

   (1) 1   (2) 2   (3) 3   (4) 4

6. The following equation represents the reaction for a zinc-copper chemical cell:

$$Zn(s) + Cu^{2+}(aq) \rightarrow Zn^{2+}(aq) + Cu(s).$$

If 0.1 mole of copper is deposited on the copper electrode, the mass of the zinc electrode will

(1) decrease by 6.5 g   (2) increase by 6.5 g
(3) decrease by 13 g   (4) increase by 13 g

7. How many moles of electrons would be required to completely reduce 1.5 moles of Al^{3+} to Al?

   (1) 0.50   (2) 1.5   (3) 3.0   (4) 4.5

8. Given the unbalanced equation:

$$Ca^0 + Al^{3+} \rightarrow Ca^{2+} + Al^0$$

When the equation is completely balanced with the smallest whole-number coefficients, what is the coefficient of Ca^0?

   (1) 1   (2) 2   (3) 3   (4) 4

PRACTICE FOR CONSTRUCTED RESPONSE

1. Use Table J of the Reference Table for Physical Setting/Chemistry to predict which of the following will react spontaneously. If there is a spontanous reaction, write the products and balance the equation. If no, write "no reaction."

   a. $Ca(s) + Ag^+(aq) \rightarrow$
      (2 points)

   b. $Pb(s) + Fe^{2+}(aq) \rightarrow$
      (2 points)

   c. $Al(s) + Fe^{2+}(aq) \rightarrow$
      (2 points)

2. Four metals, **A**, **B**, **C**, and **D**, react spontanously with their nitrate salts as shown below:

$$2A\ NO_3\ (aq)\ +\ B(s)\ \rightarrow\ B(NO_3)_2(aq)\ +\ 2A(s)$$

$$C(NO_3)_2(aq)\ +\ B(s)\ \rightarrow\ \text{no reaction}$$

$$2D\ NO_3(aq)\ +\ C(s)\ \rightarrow\ \text{no reaction}$$

Arrange these four metals into brief activity series such as shown in Table J of the *Reference Tables for Physical Setting/Chemistry.* (2 points)

3. A voltaic cell is set up below. One beaker has a copper electrode in a solution of $Cu(NO_3)_2$. The other beaker has a lead electrode in a solution of $Pb(NO_3)_2$.

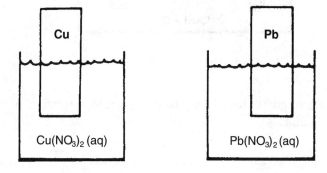

a. On the diagram above, draw and label any necessary connecting wires and the salt bridge. (2 points)

b. When connected properly, which direction will the electrons flow through the connecting wire? (1 point)

(a) toward Cu   (b) toward Pb   _____

c. Write the equation for the electrode being reduced. (1 point)

_____

d. Which electrode will be the anode? (1 point)   _____

4. A fork is connected to a battery in an electolytic cell. Fused AgCl is in the container. Silver is to be plated on the fork.

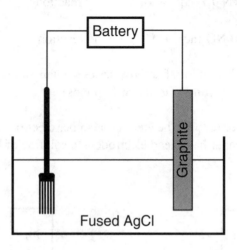

a. Must the fork be connected to the positive or negative terminal of the battery? Explain. (2 points)

_____

b. Is the fork the anode or the cathode? (1 point)

_____

c. Write the equation for the reaction at the fork. (1 point)

_____

d. Write the equation for the reaction at the graphite. (1 point)

_____

e. In which direction do electrons flow through the battery? (1 point)

_____

5. Explain the reason the following steps would slow down the corrosion of an iron bridge.

   a. Painting the bridge. (1 point)

   _____

   _____

   b. Connecting the bridge to bars of zinc. (1 point)

   _____

   _____

   c. Connecting the bridge to the negative terminal of batteries. (1 point)

   _____

   _____

CHAPTER 9
ACIDS, BASES AND SALTS

ELECTROLYTES

Acids, bases, and salts are electrolytes. Electrolytes are substances that when dissolved in water will form a solution capable of conducting electric current. In order to conduct electric current, a solution must contain ions that are free to move. The higher the concentration of ions, the better the solution conducts electric current. All ionic compounds are either acids, bases, or salts. The behavior of electrolytes was first explained by **Svante Arrhenius.**

Acids

The Arrhenius definition of an acid is a compound that yields hydrogen ions (H^+) as the only positive ion in aqueous solution. A hydrogen ion is unique among ions in that it has no electrons of any kind. It is just a proton. In aqueous solution, protons will usually stick to a water molecule forming a **hydronium** ion ($H_3O^+(aq)$). As an alternative, hydronium ion is often substituted for hydrogen ion when dealing with acids. Inorganic Arrhenius acids can be quickly identified in formulas because the first (positive) ion in the formula will be H. (HCl, HNO_3, H_2SO_4, etc.) Organic acids always end with the functional group $-COOH$. The final H is the H^+ that comes off in aqueous solution to make the solution have acidic properties. (CH_3COOH, C_2H_5COOH,etc.) Acids have distinct properties as a result of their hydrogen ions. They turn acid-base indicators characteristic colors (litmus to red), react with active metals to produce hydrogen gas, and cause the sour taste in fruit and vinegar.

Bases

The Arrhenius definition of a base is a compound that yields hydroxide ions (OH^-) as its only negative ion in aqueous solutions. Arrhenius bases are easy to identify in formulas because they always end in OH. (NaOH, KOH, $Ba(OH)_2$, etc.) Care must be taken, however, to be sure the hydroxide is ionically bonded to the positive ion. If the $-OH$ is bonded to carbon, as in CH_3OH, all bonds are covalent, and the compound is an alcohol, not a base. Bases have characteristic properties that result from their hydroxide ions. They turn acid-base indicators to distinct colors (litmus to blue), have a slippery, soapy feel, and are caustic. This can cause chemical burns to the skin.

Salts

Salts are electrolytes where hydrogen is *not* the only positive ion and hydroxide is *not* the only negative ion in aqueous solutions. Inorganic salts have formulas that do not begin with H or end with OH. ($NaCl$, KNO_3, NH_4I, $Ca(ClO_3)_2$, etc.).

Neutralization

Reactions between acids and bases to produce salts and water are referred to as neutralization reaction. The Arrhenius acid-base reaction is double replacement. The H^+ from the acid and the OH^- from the base combine to form water, HOH. The negative ion from the acid combines with the positive ion from the base to make the salt. For example:

$$HNO_3 \; + \; NH_4OH \; \rightarrow \; NH_4NO_3 \; + \; HOH$$

$$\text{Acid} \qquad\quad \text{base} \qquad\quad \text{salt} \qquad \text{water}$$

The term neutralization actually applies when equal moles of hydrogen ions and hydroxide ions are mixed, producing water.

Look at the equation for the reaction between hydrochloric acid, HCl, and sodium hydroxide, NaOH:

$$HCl \; + \; NaOH \; \rightarrow \; NaCl \; + \; H_2O$$

$$\text{acid} \qquad \text{base} \qquad\quad \text{salt} \quad\;\; \text{water}$$

The neutralization portion of this reaction occurs between hydrogen ions from the acid and hydroxide ions from the base:

$$H^+ \; + \; OH^- \; \rightarrow \; HOH \; \text{(water)}$$

The last reaction on Table I of the Reference *Tables for Physical Setting/Chemistry* shows that the neutralization reaction is **exothermic**.

Acid-base titration. The molarity of an acid (or base) of unkown concentration can be determined in the laboratory by slowly adding it to a measured volume of a base (or acid) of known concentraion until neutralization occurs. This process is known as **titration**. The equivalence point of a titration occurs when neutralization has just been attained. This **end point** can be determined by using indicators (see Chart M).

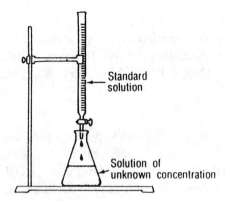

Figure 9-1. Titration.

Calculations of the molarity of unknown concentration depend on the realationship between moles and volume and on the fact that the moles of H^+ must equal the moles of OH^- at the endpoint . The moles of H^+ equals the moles of OH^- when the Molarity of the H^+ [M_{H^+}] multiplied by the volume of the H^+ [V_{H^+}] equals the Molarity of the OH^- [M_{OH^-}] multiplied by the volume of the OH^- [V_{OH^-}] .

$$M_{H^+} \times V_{H^+} = M_{OH^-} \times V_{OH^-}$$

SAMPLE PROBLEM

If 25.0mL of 2.00M HCl titrates 30.0 mL of NaOH, what is the molar concentration of the NaOH?

Solution:

1. Separate the acid data from the base data.

$M_{H^+} = 2.00M$ $\qquad\qquad\qquad$ $M_{OH^-} = ?$

$V_{H^+} = 25.0mL$ $\qquad\qquad\qquad$ $V_{OH^-} = 30.0mL$

2. Use the titration equation:

$$M_{H^+} \times V_{H^+} = M_{OH^-} \times V_{OH^-}$$

$$2.00 \tfrac{mol}{L} \times 25.0mL = M_{OH^-} \times 30.0mL$$

$$M_{OH^-} = 1.67 \tfrac{mol}{L}$$

Some acids have two or more moles of H^+ for every mole of acid. 1.0M H_2SO_4 has 2.0 $\frac{mol}{L}$ of H^+ and 1.0M of H_3PO_4 is 3.0M in H^+. The same is true for bases. While 1.0M NaOH is 1.0M in OH^-, 1.0M $Ba(OH)_2$ is 2.0M in OH^-. It is the concentration of the H^+ and OH^- that is used in the titration equation.

SAMPLE PROBLEM

If 15.0mL of 2.00M H_2SO_4 titrates 3.00M NaOH, what was the volume of the NaOH?

Solution:

1. Separate the acid data from the base data.

$$M_{ACID} = 2.00M \qquad\qquad M_{base} = 3.00M$$

$$V_{ACID} = 15.0mL \qquad\qquad V_{base} = ?$$

2. Change the concentration of the acid to the concentration of H^+ and the concentration of base to the concentration of OH^-.

$$M_{H^+} = 2 \times 2.00M = 4.00M \qquad\qquad M_{OH} = 3.00M$$

$$V_{H^+} = 15.0mL \qquad\qquad V_{OH^-} = ?$$

3. Use the titration equation.

$$M_{H^+} \times V_{H^+} = M_{OH^-} \times V_{OH^-}$$

$$4.00 \tfrac{mol}{L} \times 15.0mL = 3.00 \tfrac{mol}{L} \times V_{OH^-}$$

$$V_{OH^-} = \textbf{20.0mL}$$

Hydrogen Ion Donors and Acceptors

An alternate definion of acids and bases says that acids are hydrogen ion donors and bases are hydrogen ion acceptors.

$$HCl \rightarrow H^+ + Cl^-$$
acid

$$H^+ + NH_3 \rightarrow NH_4^+$$
base

These definitions are more general than the Arrhenius definitions. They do not require aqueous solution. For example, when the gas ammonia, NH_3, reacts with the gas hydrogen chloride, HCl, the HCl donates its H^+ to ammonia forming Cl^- and NH_4^+.

$$NH_3 + HCl \rightarrow NH_4^+ + Cl^-$$

base acid

The salt, $NH_4Cl(s)$ is produced, but not water.

QUESTIONS

1. Which substance is an electrolyte?

(1) C_2H_5OH (2) $C_6H_{12}O_6$ (3) $C_{12}H_{22}O_{11}$ (4) CH_3COOH

2. Which of the following is the best conductor of electricity?

(1) NaCl(s)   (2) NaCl (aq)   (3) $C_6H_{12}O_6(s)$ (4) $C_6H_{12}O_6(aq)$

3. Which of the following 0.1 M solutions is the best conductor of electricity?

   (1) $H_2S(aq)$ (2) HCl(aq)   (3) $C_6H_{12}O_6(aq)$ (4) $C_{12}H_{22}O_{11}(aq)$

4. Which type of reaction will occur when equal volumes of 0.1 M HCl and 0.1 M NaOH are mixed?

   (1) neutralization   (2) ionization   (3) electrolysis   (4) hydrolysis

5. The OH^- ion concentration is greater than the H_3O^+ ion concentration in a water solution of

   (1) CH_3OH (2) $Ba(OH)_2$ (3) HCl   (4) H_2SO_4

6. Which is a characteristic of an aqueous solution of HNO_3?

   (1) It conducts electricity.   (2) It forms OH^- ions.
   (3) It turns litmus blue.   (4)$[H^+]$ is less than $[OH^-]$

7. Which solution will change litmus from blue to red?

   (1) NaOH(aq)   (2) $NH_4OH(aq)$ (3) $CH_3OH(aq)$ (4) $CH_3COOH(aq)$

8. Which solution will turn litmus from red to blue?

   (1) $H_2S(aq)$ (2) $NH_4OH(aq)$ (3) $H_2SO_3(aq)$ (4) $CO_2(aq)$

9. Which substance is always produced in the reaction between hydrochloric acid and sodium hydroxide?

   (1) water   (2) hydrogen gas   (3) oxygen gas   (4) a precipitate

10. Which compound reacts with an acid to form a salt and water?

    (1) CH_3Cl (2) CH_3COOH (3) KCl   (4) KOH

11. Which equation represents a neutralization reaction?
    (1) $H^+(aq) + OH^-(aq) \rightarrow H_2O(l)$
    (2) $Ag^+(aq) + I^-(aq) \rightarrow AgI(s)$
    (3) $Zn(s) + 2HCl(aq) \rightarrow ZnCl_2(aq) + H_2(g)$
    (4) $NaCl(aq) + AgNO_3(aq) \rightarrow NaNO_3(aq) + AgCl(s)$

12. According to the Arrhenius theory, when a base is dissolved in water it will produce a solution containing only one kind of negative ion. To which ion does the theory refer?

(1) hydride   (2) hydroxide   (3) hydrogen   (4) hydronium

13. When an Arrhenius acid is dissolved in water, it produces

(1) H^+ as the only positive ion in solution
(2) NH_4^+ as the only positive ion in solution
(3) OH^- as the only negative ion in solution
(4) HCO_3^- as the only negative ion in solution

14. Which species is classified as an Arrhenius base?

(1) CH_3OH (2) $LiOH$ (3) PO_4^{3-} (4) CO_3^{2-}

15. Given the reaction:

$$HX + H_2O \rightarrow H_3O^+ (aq) + X^- (aq)$$

Based on the equation, HX would be classified as

(1) a base, because it donates a proton
(2) a base, because it accepts a proton
(3) an acid, because it donates a proton
(4) an acid, because it accepts a proton

16. Which equation illustrates H_2O acting as a hydrogen acceptor?

(1) $H^+(aq) + H_2O \rightarrow H_3O^+ (aq)$
(2) $CH_3COO^-(aq) + H_2O \rightarrow CH_3COOH (aq) + OH^- (aq)$
(3) $2Na + 2H_2O \rightarrow 2NaOH (aq) + H_2$
(4) $C + H_2O \rightarrow CO + H_2$

17. How many milliliters of 0.2 M $Ba(OH)_2$ are required to exactly neutralize 40 milliliters of 0.1 M HCl?

(1) 10   (2) 20   (3) 40   (4) 80

18. How many milliliters of a 4.0-molar solution of HCl are needed to completely neutralize 60. milliliters of a 3.2-molar solution of NaOH?

(1) 24 mL   (2) 48 mL   (3) 60. mL   (4) 75 mL

19. In the reaction $H_2O + H_2O \rightarrow H_3O^+ + OH^-$, the water is acting as

(1) a proton acceptor, only
(2) a proton donor, only
(3) both a proton acceptor and a proton donor
(4) neither a proton acceptor nor a proton donor

20. If 50. milliliters of 0.50 M HCl is used to completely neutralize 25 milliliters of KOH solution, what is the molarity of the base?

(1) 1.0 M   (2) 0.25 M   (3) 0.50 M   (4) 2.5 M

pH

All aqueous solutions have water (HOH) which ionizes into H^+ and OH^- a little bit. At room temperature, the product of the hydrogen molar concentration and the hydroxide molar concentration is 1.0×10^{-14}, a very small number. $[H^+][OH^-] = 10^{-14}$ (The [ ] means molar concentration of the substance inside.)

There will always be both H^+ and OH^- in an aqueous solution, regardless if it is an acid, base, or salt solution.

In acids, $[H^+]$ is greater than $[OH^-]$

In bases, $[H^+]$ is less than $[OH^-]$

When neutral $[H^+]$ is equal to $[OH^-] = 10^{-7}$ **M**

The pH scale was devised to easily show how acidic or alkaline a solution is. The lower the pH, the greater the $[H^+]$ and the lower the $[OH^-]$ so the more acidic the solution. A decrease of one pH unit is a 10 times increase in $[H^+]$. The pH is 7 when the $[H^+]$ equals the $[OH^-]$ at 10^{-7} **M.** The higher the pH is above 7, the lower the $[H^+]$, the higher the $[OH^-]$, and the more basic (alkaline) the solution. The following chart illustrates the relationship between pH, $[H^+]$, $[OH^-]$, acidity, and alkalinity.

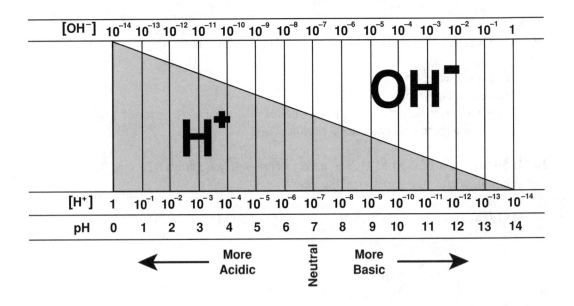

Figure 9-1. The pH chart.

QUESTIONS

Answer the following questions using Figure 9-1 above.

1. What is the pH of a solution whose H⁺ ion concentration is 0.0001 mole per liter?

   (1) 1   (2) 10   (3) 14   (4) 4

2. What is the concentration of H⁺ ions, in moles per liter, of a 0.0001 **M** HCl solution?

   (1) 1×10^{-1} (2) 1×10^{-2} (3) 1×10^{-3} (4) 1×10^{-4}

3. What is the pH of a solution that has a hydrogen ion concentration of 1×10^{-10} mole per liter?

   (1) 1   (2) 10   (3) 14   (4) 4

4. What is the hydroxide ion concentration of a solution that has a hydronium ion concentration of 1×10^{-9} mole per liter at 298 K?

(1) 1×10^{-5} mole per liter   (2) 1×10^{-7} mole per liter
(3) 1×10^{-9} mole per liter   (4) 1×10^{-14} mole per liter

5. Which concentration indicates a basic solution at 298K?

(1) $[OH^-] > 1.0 \times 10^{-7}$ (2) $[OH^-] = 1.0 \times 10^{-7}$
(3) $[H_3O^+] > 1.0 \times 10^{-7}$ (4) $[H_3O^+] = 1.0 \times 10^{-7}$

6. What is the H^+ ion concentration of an aqueous solution that has a pH of 11?

(1) 1.0×10^{-11} mol/L   (2) 1.0×10^{-3} mol/L
(3) (1) 3.0×10^{-1} mol/L   (4) 11×10^{-1} mol/L

7. If a solution has a hydronium ion concentration of 1×10^{-9} M, the solution is

(1) basic and has a pH of 9   (2) basic and has a pH of 5
(3) acidic and has a pH of 9   (4) acidic and has a pH of 5

8. When equal volumes of 0.5 **M** HCl and 0.5 **M** NaOH are mixed, the pH of the resulting solution is

(1) 1   (2) 2   (3) 7   (4) 4

9. Adding 0.1 **M** NaOH to a 0.1 **M** solution of HCl will cause the pH of the solution to

(1) decrease   (2) increase   (3) remain the same

10. The $[H^+]$ of a solution is 1×10^{-2} at 298 K. What is the $[OH^-]$ of this solution?

(1) 1×10^{-14} (2) 1×10^{-12} (3) 1×10^{-7} (4) 1×10^{-2}

11. As a solution of NaOH is diluted from 0.1 **M** to 0.001 **M**, the pH of the solution

(1) decreases   (2) increases   (3) remains the same

12. Which 0.1M solution has the highest concentration of H_3O^+ ions?

(1) CH_3COOH (2) NaCl   (3) KBr   (4) $Ba(OH)_2$

13. Which could be the pH of a solution whose H^+ ion concentration is less than the OH^- ion concentration?

(1) 9   (2) 2   (3) 3   (4) 4

PRACTICE FOR CONSTRUCTED RESPONSE

1. Identify each of the following as being (1) acid (2) base (3) salt (4) non-electrolyte. (1 point each)

   a. H_2CrO_4 _____

   b. NaCl _____

   c. H I _____

   d. Li OH _____

   e. $Ba(OH)_2$ _____

   f. $Ca(NO_3)_2$ _____

   g. C_2H_4 _____

   h. CH_3OH _____

   i. CH_3COOH _____

2. 25.0 mL of 0.350 **M** HCl(aq) is titrated to the end point with 0.535 **M** KOH(aq).

   a. Calculate the volume of KOH(aq) used. Show your work. (2 points)

   _____

   b. Explain how the answer would be different if 0.535 **M** $Ba(OH)_2$ had been used in the titration instead of KOH. (1 point)

   _____

   _____

3. When solutions of sulfuric acid and potassium hydroxide are mixed:

   a. write the balanced equation for the resulting reaction. (2 points)

   _____

   b. Will this reaction be endothermic or exothermic? (1 point)

   _____

4. A solution is thought to be very acidic. Which indicator on Table M of the Reference Tables for Physical Setting/Chemistry would be the best choice to determine if it is indeed very acidic? Explain your answer. (2 points)

_____

_____

_____

_____

5. A student has a solution in the chemistry laboratory. Give two different safe methods the student can use to determine if it is acidic, basic, or neutral. (2 points)

_____

_____

_____

_____

CHAPTER 10
NUCLEAR CHEMISTRY

NATURAL RADIOACTIVITY

Elements that naturally emit energy without the absorption of energy from an outside source are said to be "naturally radioactive." Elements with atomic number greater than 83 (bismuth) have no known stable isotopes. They are all radioactive. Radioactivity is associated with some form of disintegration, or "decay," of the nucleus. This disintegration results in the emission of particles and/or radiant energy.

While most isotopes are stable and not radioactive, those that are radioactive have an unstable ratio of protons and neutrons in their nuclei. A nucleus becomes more stable by throwing out a piece of the nucleus (the emissions) until the ratio becomes stable.

When the nucleus of an element disintegrates, it changes to the nucleus of another element. This change from one element to another is called **transmutation.**

Types of Emissions

Soon after the discovery of radioactivity, three different types of emissions were identified. Identification was made on the basis of differences in behavior in an electric field. Those emissions that were attracted to the negative electrode were labeled *alpha* beams. Those attracted to the positive electrode were called *beta* beams. Emissions that were apparently unaffected by charge were called *gamma* beams. It was later found that these emissions also differed from one another in mass, penetrating power, and ionizing power. These findings enabled scientists to identify those three different types of emissions.

Since those early discoveries, other types of emissions have been found. One of those has the mass of beta particles, but it is attracted to the negative electrode. It is called a *positron*.

Alpha Decay. Alpha particles have been identified as helium nuclei (2 protons and 2 neutrons) that are ejected at high speeds from certain radioactive isotopes. The reaction involved in nuclear disintegration resulting in the emission of alpha particles is called *alpha decay*. The emitting atoms are called *alpha emitters*. The symbols used to designate isotopes, particles, or radiations in nuclear reactions follow this pattern:

$$_{Z}^{A}X$$

In which X is the symbol of the atom or emission, A is the mass number (numbers of protons + neutrons), and Z is the atomic number (number of protons). See page: 2. The atomic number is also the charge of the particle.

The alpha particle is designated:

$$_{2}^{4}\text{He}$$

An example of an equation for an alpha decay would be:

$$_{88}^{226}\text{Ra} \rightarrow {_{86}^{222}}\text{Rn} + {_{2}^{4}}\text{He}$$

Notice that in nuclear reactions, both mass and charge are balanced. The total mass numbers of the reactants equals the total mass numbers of the products. Also, the total atomic numbers of the reactants equals those of the products.

In alpha emission, the isotope's atomic number is reduced by *two,* and its mass number is reduced by *four.*

Beta Decay. Beta particles have the same charge and mass as electrons. Therefore, they are described as electrons ejected at high speeds from certain radioactive isotopes. The symbol for a beta particle is:

$$_{-1}^{0}\text{e}$$

The reaction is called *beta decay,* and the atom is called a *beta emitter.* The equation for the emission of a beta particle from $_{92}^{235}\text{U}$ would be written:

$$_{92}^{235}\text{U} \rightarrow {_{-1}^{0}}\text{e} + {_{93}^{235}}\text{Np}$$

Since a beta particle has a charge of minus one, balancing the charges brings about an *increase* of one in the atomic number of the product isotope. Beta decay raises at least two questions:

1. How can the product gain a proton without gaining mass?
2. How can the nucleus emit an electron it does not have?

The accepted explanation involves a series of reactions, called *neutron decay,* which results in a neutron in the nucleus being transformed into a proton with the "creation" of an electron.

$$n \rightarrow p + e^{-}$$

Positron Emissions. Positron particles have the same mass as electrons, but they have a positive charge. The are known as anti-electrons. When positrons meet electrons, they annihilate each other producing large amount of energy (gamma radiation). This happens almost immediately when positrons meet matter.

$$_{+1}^{0}e \ + \ _{-1}^{0}e \ \rightarrow \ _{0}^{0}\gamma$$

The equation for the emission of a positron from sodium–22 would be written:

$$_{11}^{22}Na \rightarrow \ _{+1}^{0}e \ + \ _{10}^{22}Ne$$

Positrons come from the decay of protons in the nucleus into a positron and a neutron.

$$_{1}^{1}p \ \rightarrow \ _{+1}^{0}e \ + \ _{0}^{1}n$$

Mass is conserved because, although a proton is lost, a neutron is gained. They both have a mass of one atomic mass unit. The proton and positron are both charged +1.

Gamma Radiation. Gamma rays are a high energy form of light energy. Since gamma rays are not affected by the electric field, they are said to possess zero charge. Since they have no mass, they are not considered particles. Because of their penetrating ability, their properties resemble those of X-rays, but they have much higher energies than most X-rays.

Characteristics of Alpha, Beta and Gamma Emissions

Name	Symbol	Charge	Mass (amu)	Relative Penetrating Power	Relative Ionizing Power
Alpha	$_{2}^{4}He$	+2	4	low	10,000
Beta	$_{-1}^{0}e$	−1	0	moderate	100
Gamma	$_{0}^{0}\gamma$	0	0	high	1
positron	$_{+1}^{0}e$	+1	0	annihilated upon meeting electrons	

Detection of Radioactivity

The very nature of radioactivity lends itself to various methods of detection and study, both quantitative and qualitative. This "nature" includes its ionizing, fluorescent, and photographic effects.

Half-Life

Radioactive isotopes disintegrate at characteristic rates. The time interval required for half the sample to disintegrate is called *half - life* of that isotope. The half-life of iodine-131 ($^{131}_{53}$I) is 8 days. If we started with a 100-gram sample of I-131, after 8 days only 50 grams of iodine-131 would remain. After *another* 8 days had passed, only half of the 50 grams, or 25 grams, would remain. After another 8 days, only 12.5 grams would remain and so on.

Although formulas have been derived for calculating the mass of a radioactive isotope remaining after a given time, it is convenient to set up a simple table of mass and time. For example in the case of iodine-131, how much of a 100-gram sample will remain after 24 days?

Time	Mass
0 days	100g
8 days	50g
16 days	25g
24 days	12.5g

Remember, it is important to:

1. Find the half-life in Reference Table N, if not given.
2. Begin your table with zero time.

QUESTIONS

Answer the following questions using Tables N and O of the *Reference Tables for Physical Setting/Chemistry.*

1. Which element has no known stable isotope?

(1) carbon

(2) potassium

(3) polonium

(4) phosphorus

2. Which type of radiation, when passed between two electrically charged plates, would be deflected toward the positive plate?

(1) an alpha particle   (2) a beta particle
(3) a neutron   (4) a positron

3. In an electric field, which emanation is deflected toward the negative electrode?

(1) beta particle   (2) alpha particle
(3) x rays   (4) gamma rays

4. Which particle is given off when $^{32}_{15}P$ undergoes a transmutation reaction?

(1) an alpha particle   (2) a beta particle
(3) a neutron   (4) a positron

5. If 8.0 grams of a sample of ^{60}Co existed in 1990, in which year will the remaining amount of ^{60}Co in the sample be 0.50 gram?

(1) 1995   (2) 2000   (3) 2006   (4) 2011

6. What is the number of hours required for potassium-42 to undergo 3 half-life periods?

(1) 6.2 hours   (2) 12.4 hours
(3) 24.8 hours   (4) 37.2 hours

7. In the equation $^{234}_{91}Pa \rightarrow {}^{234}_{92}U + X$ the X represents a

(1) helium nucleus   (2) beta particle   (3) proton   (4) neutron

8. In the equation $^{232}_{90}Th \rightarrow {}^{228}_{88}Ra + X$ which particle is represented by the letter X?

(1) an alpha particle   (2) a beta particle
(3) a positron   (4) a deuteron

9. Given the reaction:

$$^{234}_{91}Pa \rightarrow X + {}^{0}_{-1}e$$

When the equation is correctly balanced, the nucleus represented by X is

(1) $^{234}_{92}U$ (2) $^{235}_{92}U$ (3) $^{230}_{90}Th$ (4) $^{232}_{90}Th$

10. As an atom of a radioactive isotope emits an alpha particle, the mass number of the atom

(1) decreases   (2) increases   (3) remains the same

11. In the natural radioactive decay of K-37, which of the following is the decay product?

(1) Cl-33   (2) Sc-41   (3) Ar-37   (4) Ca-37

ARTIFICIAL RADIOACTIVITY

Natural radioactivity is the decay process occurring in elements that are radioactive as they exist in nature. Radioactive elements can also be produced by bombarding the nuclei of stable atoms with high energy particles, such as protons, neutrons, and alpha particles. Such radioactivity is called **artificial radioactivity**. An example of artificial radioactivity is the nuclear reaction that takes place when the element beryllium is bombarded by protons.

$$_{4}^{9}\text{Be} + _{1}^{1}\text{H} \rightarrow _{3}^{6}\text{Li} + _{2}^{4}\text{He}$$

Artificial Transmutation

When the nuclei of stable atoms are bombarded by accelerated particles, the nuclei become unstable and may cause the formation of isotopes of new elements. This process is called **artificial transmutation**. When this process occurs as the result of natural radioactivity, it is called **natural transmutation**. An example of artificial transmutation is the bombardment of the stable isotope Al-27 with an alpha particle:

$$_{13}^{27}\text{Al} + _{2}^{4}\text{He} \rightarrow _{15}^{30}\text{P} + _{0}^{1}\text{n}$$

Accelerators. Particle accelerators of various types are used to give charged particles enough kinetic energy to overcome electrostatic forces and thus penetrate nuclei of target atoms. Acceleration is accomplished by means of the manipulation of electric and magnetic fields.

NUCLEAR ENERGY

Nuclear reactions involve energies that are millions of times greater than those found in ordinary chemical reactions. Energies of these magnitudes often are the result of the conversion of mass into energy.

Fission Reaction

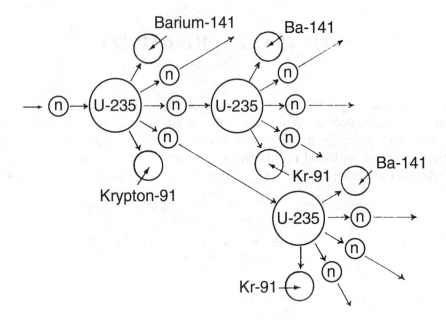

Figure 10-1. Nuclear Fission

Fission is the splitting of a heavy nucleus into two lighter nuclei. Only elements of high atomic number can be fissioned. The fission reaction is brought about by the capture of neutrons. This results in the liberation of energy and the release of two or more neutrons per atom, in addition to the fission fragments. The energy liberated results from the conversion of mass into energy. The fact that each fission process emits more than one neutron is very important. The emitted neutrons induce other nuclei to split. In nuclear reactors, the *chain reaction* that results is controlled. In an atomic bomb, it is not controlled.

When heavy elements undergo fission, the new elements formed are more stable than the parent element because of the greater binding energy per nucleon.

Fission reactions produce the energy in nuclear reactors to make electricity. The fuel will last longer than fossil fuels. The carbon dioxide and smoke products of fossil fuels are not produced by fission so nuclear reactors do not contribute to air pollution, global warming, or acid rain. Fission reactions also produce the energy in some nuclear weapons.

A problem with fission reactions is that both the fuels and products are radioactive which poses a health hazard if people are exposed. Other worrisome problems are the possibility of nuclear accident or the possible theft of fuels from reactor sites (to produce weapons).

Radioactive Wastes

Fission products from nuclear reactors are highly radioactive and remain so for considerable periods of time. These highly dangerous substances cannot simply be discarded. Solid and liquid wastes, such as strontium-90 and cesium-137 are sealed in special containers that are stored underground or in isolated areas. Low-level radioactive wastes may be diluted and released into the environment. Gaseous radioactive wastes, such as radon-222, krypton-85, and nitrogen-16 are stored until they decay to safe levels, and then dispersed into the atmosphere.

Fusion Reaction

When two or more light nuclei combine to form a single nucleus of greater mass, the reaction is called *nuclear fusion,* or a *fusion reaction.* The energy released in fusion reactions is much greater than that in fission reactions. The mass of the new nucleus is *less* than the sum of the masses of the light nuclei. The difference in mass represents the mass that was converted to energy in the process. Some of this energy provides for the greater binding energy per nucleon and, therefore, the greater stability of the heavier and more stable nucleus formed.

Solar energy is believed to be the result of the fusion of ordinary hydrogen atoms into helium atoms. In order to achieve the energy necessary to initiate the fusion reaction of a hydrogen bomb, it is necessary to use a fission reaction as a supplier of heat and pressure.

Fuels. The fuels utilized for fusion reactions are the hydrogen isotopes:

$$\text{deuterium, } {}^{2}_{1}\text{H and tritium, } {}^{3}_{1}\text{H.}$$

Deuterium may be obtained from heavy water, which can be extracted from ordinary water. Tritium is manufactured by the nuclear reaction:

$$ {}^{6}_{3}\text{Li} + {}^{1}_{0}\text{n} \rightarrow {}^{3}_{1}\text{H} + {}^{4}_{2}\text{He} $$

High energy requirement. In order for nuclei to combine, they must have sufficient energy to overcome the forces of repulsion that exist between particles of like charge (protons). The magnitude of the repulsive force increases with the size of the charge. Thus, only small nuclei, with very small charge, can be used in fusion reactions. Fusion with ordinary hydrogen, ${}^{1}_{1}\text{H}$ is very slow. More rapid fusion reactions occur with the heavier hydrogen isotopes.

The idea of obtaining energy from fusion reactions is a very attractive one. Fusion reactors would use inexpensive, plentiful fuel and would produce practically no harmful radioactive wastes. However, there are some major problems that must be solved before practical fusion energy is a reality. These problems involve the extremely high activation energies necessary to trigger the fusion reaction. In order for fusion reactions to occur, temperatures on the order of 10^9 degrees C are required. The problems that must be solved are: How will these temperatures be achieved, and how will they be contained? The search for answers to these questions continues.

USES OF RADIOISOTOPES

Major uses of radioactive isotopes can be classified into three basic groups: (1) uses based on chemical reactivity   (2) uses based on radioactivity   (3) uses based on half-life.

Based on chemical reactivity. Fortunately, the chemical behavior of radioactive isotopes is the same as that of the stable isotopes of the same elements. As a result, radioactive isotopes can be used to trace the course of a reaction. The reaction continues in a normal fashion, but the path of the substituted radioactive material can be traced using radioactivity detection devices and methods. Many organic reaction mechanisms, including those in living systems, are studied using carbon-14 as the tracer.

Based on radioactivity. Isotopes with very short half-lives are administered to patients for diagnostic purposes. Tumors in various organs can be located and levels of activity of those organs can be monitored by administering and tracing radioactive substances known to concentrate in those organs. Technetium-99 is used to determine the location of brain tumors. Iodine-131 is used for the diagnosis of thyroid disturbances.

Cancer cells are more sensitive to radiation than are normal cells. Using carefully selected dosages, radioisotopes have been successfully used to treat cancer by destroying malignant cells. Dosages are sought that will attack only the malignant cells and not the normal cells. Radium and cobalt-60 are used in cancer therapy.

Radiation can be used to destroy bacteria, yeasts, and molds, as well as the eggs of insects. These capabilities provide the bases for the use of radioisotopes in food preservation.

Since the decrease in intensity of radiation is dependent on the amount of matter it passes through, radiation is used to measure thicknesses of industrial products. The wear-resistant properties of the moving parts of an engine may be measured by using radioactive parts and measuring the amount of radiation transferred to the lubricating materials used.

Based on half-life. Radiochemical dating is a method of determining the age of fossils and rocks. The ratio of uranium-238 to lead-206 in a mineral can be used to determine the age of the mineral.

The half-life of a radioisotope is a constant factor. It is not affected by temperature, pressure, or any external factors. Thus, it can be assumed that a radioactive substance is decaying at the same rate today that it did at the time of its origin. The use of carbon-14 measurements in dating recent fossils relies on the fact that the activity of this isotope in living organisms is the same as it is in the atmosphere. It is assumed that the level of carbon-14 activity was the same when the fossil formed as it is today.

As long as an organism is living, the ratio of carbon-14 to carbon-12 is constant and remains constant. As soon as the organism dies, the carbon-14 lost through radioactive decay is not replaced. Thus, the ratio of the carbon-14 present in a fossil to that in the atmosphere can be used to determine the age of the fossil.

QUESTIONS

Answer the following questions using Tables N and O of the *Reference Tables for Physical Setting/Chemistry*.

1. A positively charged particle has great difficulty penetrating a target nucleus because the charged nucleus has

   (1) a positive charge, which repels the particle
   (2) a negative charge, which attracts the particle
   (3) the protection of surrounding electrons
   (4) a very high binding energy

2. Which radioactive waste can be stored for decay and then safely released directly into the environment?

   (1) N-16   (2) Sr-90   (3) Cs-137   (4) Pu-239

3. Which equation represents a nuclear reaction that is an example of artificial transmutation?

   (1) $^{43}_{21}Sc \rightarrow\ ^{43}_{20}Ca +\ ^{0}_{+1}e$ (2) $^{14}_{7}N +\ ^{4}_{2}He \rightarrow\ ^{17}_{8}O +\ ^{1}_{1}H$

   (3) $^{10}_{4}Be \rightarrow\ ^{10}_{5}B +\ ^{0}_{-1}H$ (4) $^{14}_{6}C \rightarrow\ ^{14}_{7}N +\ ^{0}_{-1}e$

4. The operation of a commercial nuclear reactor in New York State requires an isotope that will undergo

   (1) fission and a controlled chain reaction
   (2) fission and an uncontrolled chain reaction
   (3) fusion and a controlled chain reaction
   (4) fusion and an uncontrolled chain reaction

5. Which particle can *not* be accelerated by the electric or the magnetic field in a particle accelerator?

   (1) electron   (2) neutron   (3) helium nucleus   (4) hydrogen nucleus

6. A radioactive-dating procedure to determine the age of a mineral compares the mineral's remaining amounts of isotope ^{238}U and isotope

   (1) ^{206}Pb (2) ^{206}Bi (3) ^{214}Pb (4) ^{214}Bi

7. Particle accelerators can be used to increase the kinetic energy of

   (1) deuterium   (2) neutrons   (3) protons   (4) tritium

8. Given the reaction $^{7}_{3}Li + X \rightarrow \ ^{8}_{4}Be$

   Which species is represented by X ?

   (1) $^{1}_{1}H$ (2) $^{2}_{1}H$ (3) $^{3}_{2}He$ (4) $^{4}_{2}He$

9. Radioisotopes used for medical diagnosis must have

   (1) long half-lives and be quickly eliminated by the body
   (2) long half-lives and be slowly eliminated by the body
   (3) short half-lives and be quickly eliminated by the body
   (4) short half-lives and be slowly eliminated by the body

10. Compared to an ordinary chemical reaction, a fission reaction will

   (1) release smaller amounts of energy
   (2) release larger amounts of energy
   (3) absorb smaller amounts of energy
   (4) absorb larger amounts of energy

11. Given the equation: $^{14}_{7}N + \ ^{4}_{2}He \rightarrow X + \ ^{17}_{8}O$

   When the equation is correctly balanced, the particle represented by the X will be

   (1) $^{0}_{-1}e$ (2) $^{1}_{0}n$ (3) $^{1}_{1}H$ (4) $^{2}_{1}H$

12. Iodine-131 is used for diagnosing thyroid disorders because it is absorbed by the thyroid gland and

(1) has a very short half-life   (2) has a very long half-life
(3) emits alpha radiation   (4) emits gamma radiation

13. Given the nuclear reaction: $^{12}_{6}C + ^{2}_{1}H \rightarrow X + ^{1}_{0}n$

When the equation is correctly balanced, the nucleus represented by the X is

(1) $^{14}_{6}N$ (2) $^{13}_{7}N$ (3) $^{13}_{7}C$ (4) $^{13}_{6}C$

14. Which radioactive isotope is often used as a tracer to study organic reaction mechanisms?

(1) carbon -12 (2) carbon -14 (3) uranium -235 (4) uranium -238

15. In the reaction $^{27}_{13}Al + ^{4}_{2}He \rightarrow X + ^{1}_{0}n$ the isotope represent by X is

(1) $^{29}_{12}Mg$ (2 $^{28}_{13}Al$ (3) $^{27}_{14}Si$ (4) $^{30}_{15}P$

PRACTICE FOR CONSTRUCTED RESPONSE

1. Below are four nuclear reactions:

(1) $^{3}_{2}He + ^{1}_{1}H \rightarrow ^{4}_{2}He +$ _____

(2) $^{1}_{0}n + ^{235}_{92}U \rightarrow ^{139}_{56}Ba + 3^{1}_{0}n +$ _____

(3) $^{14}_{6}C \rightarrow ^{0}_{-1}e +$ _____

(4) $^{226}_{88}Ra \rightarrow ^{222}_{86}Rn +$ _____

    a. Complete each balanced equation with a single product.
       (4 points)

    b. Which reaction is alpha decay?   _____
       (1 point)

    c. Which reaction is fission?   _____
       (1 point)

    d. Which reaction is fusion?   _____
       (1 point)

    e. Which reaction is beta decay?   _____
       (1 point)

2. Complete the balanced nuclear equations for the natural radioactive decay of the following nuclides:
(6 points)

   a. $^{220}_{87}\text{Fr} \rightarrow$ _____

   b. $^{16}_{7}\text{N} \rightarrow$ _____

   c. $^{37}_{20}\text{Ca} \rightarrow$ _____

3. The exposure to and uses of radioactive isotopes is a controversial topic.

   a. List the benefits of radioactivity to humans. (2 points)

   _____

   _____

   b. List the risks of radioactivity to humans. (2 points)

   _____

   _____

4. Nuclear fusion produces enormous energy. (The Sun and other stars are fueled by fusion.) Fusion fuel is abundant and inexpensive. (The ocean is full of hydrogen). Fusion fuel and products are not radioactive. But, all working nuclear reactors in the world use fission reactions which have many risk factors. What makes fusion reactors difficult to make? (2 points)

_____

_____

_____

_____

_____

_____

5. A paint sample of an "old painting" is found to contain 0.010 g of $^{137}_{55}Cs$ (a radioactive isotope) and 0.030 g of $^{137}_{56}Ba$ (not radioactive).

a. Write the nuclear equation for the decay of ^{137}Cs. (2 points)

_____

b. How old is the paint? (1 point)

_____

c. What assumption must be made in order to use this method of dating materials? (1 point)

_____

_____

APPENDIX A

USEFUL MATHEMATICS IN CHEMISTRY

Scientific Notation

It is convenient to write very large and very small numbers in exponential form as a way to keep track of the size of the number. For example:

Size	Number	Equivilent number	Scientific Notation
hundreds	$255 =$	$2.55 \times 100 = 2.55 \times (10^1 \times 10^1)$	$= 2.55 \times 10^2$
thousands	$3666 =$	$3.666 \times 1000 = 3.666 \times (10^1 \times 10^1 \times 10^1) =$	3.666×10^3
tenths	$0.255 =$	$\frac{2.55}{10} = \frac{2.55}{10^1}$	$= 2.55 \times 10^{-1}$
hundredths	$0.0255 =$	$\frac{2.55}{100} = \frac{2.55}{10^2}$	$= 2.55 \times 10^{-2}$
thousandths	$0.00266 =$	$\frac{2.55}{1000} = \frac{2.55}{10^3}$	$= 2.55 \times 10^{-3}$

Although 255, 25.5×10^1, and 2.55×10^2 are all the same number, only 2.55×10^2 is in scientific notation because the 2.55 is between 1 and 10 (but not including 10). By looking at the 10^2 the number's size is immediately recognized to be in the hundreds.

SIGNIFICANT FIGURES

Many operations in the chemistry laboratory involve measurements. Finding mass and volume, for example, both require measurement operations. Recording the results of these measurements should be based on certain rules in order to show the precision of the measurement.

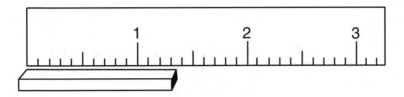

Figure 12-1

All measurements involve uncertainty which is ordinarily recorded in the last digit. For example, the length of the object shown being measured is between 1.3 and 1.4 units. Its length may be recorded as 1.35 units. The last digit, 5, is estimated, but it is considered to be significant and should be recorded. In interpreting a recorded mass of 4.7 grams, the assumption would be that the object was massed to the nearest tenth of a gram and that its exact mass is between 4.65 grams and 4.75 grams. If the recorded mass of the object were 4.733 g, there are four significant figures, It must be assumed that the object was massed to the nearest thousandth of a gram. Significant digits. consist of those that we know with certainty plus one more doubtful, or estimated, digit.

Zeros may or may not be significant, according to a set of conventional rules. The number 6.00 has three significant figures, the two zeros being significant. This number implies that the measurement was made to the nearest hundredth. The zeros in 4.023, 6.105, 8.210, 20.10, and 3.002 are all considered significant. Each of these numbers contains four significant figures. One should always record all quantities provided by the measurement device used. An object massed as exactly two grams on a balance that is accurate to the nearest hundredth of a gram should be recorded as 2.00g, not 2g.

When a zero is used to locate a decimal point, as in 0.023, or 4500, the zeros are *not* significant figures. The numbers 0.036, 87000, and 0.0020 have only *two* significant figures each. These numbers could be written in scientific notation, 3.6×10^{-2}, 8.7×10^{4}, and 2.0×10^{-3}. All numbers in scientific notation are significant. This is one of the advantages of the use of scientific notation.

Calculations and Significant Figures

In any calculation in which experimental results are used, the final result should contain only as many significant figures as are justified by the instruments employed in finding those results. The measurement of *least* precision determines the number of significant figures in the final answer.

Addition and subtraction. In addition and subtraction, retain only as many *decimal places* in the result as there are in that item which uses the *least* number of decimal places.

Example:

$$
\begin{array}{r}
42.2 \text{ (limiting measurement)} \\
3.024 \\
+ \ 7.23 \\
\end{array}
$$

Answer: 52.5

Multiplication and division. In multiplication and division problems, the answer should contain only as many significant figures as are contained in the item with the *least* number of significant figures.

Example: $43.42 \times 0.029 \times 44.6 = 56$

Since the number 0.029 has only two significant figures, the answer must be 56, which contains two significant figures.

Percent Error

All experiments involve some error. Consider, for example, a problem based on the "experimental data," the molar volume of hydrogen gas at STP is determined to be 22 100 milliliters. The accepted value for the molar volume of any gas at STP is 22 400 milliliters. The error in this instance is quite small.

The percent error of any given or measured value is a comparison of that value with the number that is accepted as being the true value. The following expression can be used ito find percent error:

$$
\% \text{ error} = \frac{\text{difference between accepted and experimental values}}{\text{accepted value}} \times 100\%
$$

The percent error for the sample problem data is:

$$
\frac{22\ 400\ \text{mL} - 22\ 100\ \text{mL}}{22\ 400\ \text{mL}} = \times 100\% = 1\%
$$

Graphing

A graph is an often revealing way to display the data of two variables. If all other variables are kept constant, the relationship between the two variables can be determined.

Consider the following graphs with their table of values:

1.

y	x
4	1
4	2
4	3
4	4

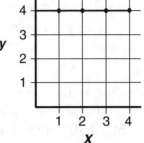

In **graph 1**, y does *not* depend on x. No matter what value of x is chosen, y stays the same constant value.

$$y = \text{constant}$$
$$y = 4$$

x is not even in the equation.

Graphs 2 and 3 are both direct relationships. When x gets bigger, so does y. They are also both *linear* (form a straight line). Linear graphs fit the linear eqation:

$$y = mx + b$$

$$m = \text{slope of graph} = \frac{\text{rise}}{\text{run}} = \frac{\Delta y}{\Delta x}$$

$$b = y \ \text{intercept}$$

2.

y	x
3	1
5	2
7	3
9	4

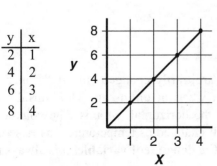

Graph 2,

$$m = \frac{\Delta y}{\Delta x} = \frac{5\text{-}3}{2\text{-}1} = 2 \text{ and } b = 1$$

The equation for **graph 2** is
$$y = 2x + 1$$

3.

y	x
2	1
4	2
6	3
8	4

Graph 3 is also linear where

$$m = \frac{\Delta y}{\Delta x} = \frac{4\text{-}2}{2\text{-}1} = 2 \ \text{But } b = 0 \text{ because}$$

the graph goes through the origin. The equation for graph 3 is:

$$y = 2x \ \text{ or } \ \frac{y}{x} = 2.$$ This is the equation of a direct proportion. If x doubles, the y doubles, etc.

Examples of direct proportions are found between the volumes, temperatures, and pressures of gases.

4.

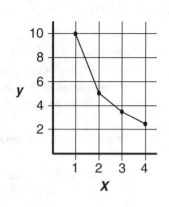

y	x
10.	1
5.0	2
3.3	3
2.5	4

$\dfrac{V}{T}$ = constant and $\dfrac{P}{T}$ = constant.

(as long as the temperature is in the absolute or Kelvin scale).

Graph 4 shows an indirect or inverse relationship between x and y. When x gets bigger, y gets smaller. In this case x • y = a constant value, 10. The equation for the graph is :

$$x \cdot y = 10$$

This is the equation of an inverse proportion. If x doubles, y becomes one half, etc. An example of an inverse proportion is the volume of a constant amount of gas and its pressure, at constant temperature.

$$P \cdot V = \text{constant}$$

Suppose a constant amount of gas was heated in l0.°C increments while the pressure was kept constant. The resulting volume of the gas was measured giving the following data for the temperatures used:

T (°C)	V (mL)
20.	1230
30.	1270
40.	1310
50.	1360
60.	1400

In this experiment, the temperature was arbitrarily changed in 10.°C increments. Different intervals could have been used. It was not unknown. This variable is the **independent variable.** It is always plotted on the horizontal (x) axis. The volume was a measurement whose value we did not choose. It changed as temperature was raised. Since its value depends on the temperature used, it is **the dependent variable.** It is always plotted on the vertical (y) axis.

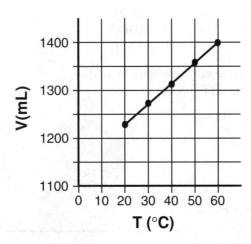

The graph shows the volume is directly related to Celcius temperature, and it is linear. It is not, however, directly proportional because the graph does not go through the origin. It fits the general linear equation y = mx + b. Here :

m = 1270-1230 mL / 30-20 °C = 4.0 mL/ °C and b = 1140 mL.

Therefore, V = 4.0 mL/°C T + 1140 mL is the equation.

Points beyond those temperatures where we have data can be found by **extrapolation.** It is risky to extrapolate too far beyond the data because it assumes the relationship continues, and we have no data to verify that. We had to extrapolate to find b in our equation because there was no data for T = 0 °C. To find the temperature where the volume shrinks to 0 mL takes us much farther from our data. Assuming the relationship holds, substitute 0 into the equation for V, and solve for T. The answer is -285°C which should be absolute zero, -273°C. There is apparently some experimental, error.

The percent error is -273-(-285) °C / -273 °C ×100%= 4.4 % error.

It is usually more reliable to find answers between our data points by **interpolation.** By looking at our line, the volume when the temperature is 25°C must be 1250 mL.

QUESTIONS

1. Write the number 24388 in scientific notation. 1. _____

2. Write the number 0.0036 in scientific notation. 2. _____

3. How many significant figures are in each of the following numbers:

   a. 234 _____   d. 12.00 _____

   b. 2500 _____   e. 0.01010 _____

   c. 0.0034 _____   f. 2.5×10^4 _____

4. Give the answer to the following calculations in the proper number of significant figures.

   a. 25,30 cm × 0.035 cm =   4.a _____

   b.   25.30 cm   b. _____
          123.0 cm
       +   1.33 cm

5. The results of an experiment shows a heat of fusion to be 5.56 kJ/g, but the accepted value is 4.95 kJ/g. Calculate the percent error.

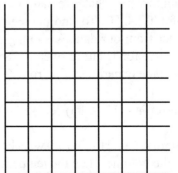

5. _____

6. Consider the following data:

T(°C)	P(kPa)
10.	722
20.	747
30.	773
40.	798

6.a _____

   a. Is temperature or pressure most likely the independent variable?

   b. Plot the graph on the axes provided below. Lable the axes.   b _____

   c. Is P(kPa) directly related to T(°C)?   c _____

   d. Is P(kPa) directly proportional to T(°C)?   d _____

   e. What is the value for the slope of the line?   e _____

   f. What is the value for P when T is 0°C?   f _____

   g. What is the value from this graph for T when P is 0 kPa?   g _____

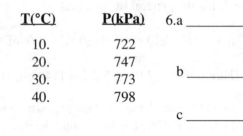

   h. Write the equation relating P and T.   h _____

APPENDIX B

NOTABLE SCIENTISTS AND THEIR
CONTRIBUTIONS IN CHEMISTRY

Robert Boyle 1627-1691   Boyle was born in Ireland to English nobility. He was a wealthy scholar who helped found the Philosophical College, the predecessor of the Royal Society of London. Boyle was one of the first to use the scientific method while experimenting. He is best known for his work with gases, discovering the inverse proportion between the pressure and volume at constant temperature. $P \times V = constant$

Antoine Lavoisier 1743-1794   Lavoisier was a French chemist and member of the Academy of Sciences in Paris.   He is often thought of as the founder of modem chemistry. Lavoisier disproved the theory of phlogiston by quantitatively showing that oxygen in the air combines with a substance being burned. He provided experimental evidence for the law of conservation of mass. He was guillotined during the French revolution.

John Dalton 1766-1844 Dalton was an English chemist and teacher. He was interested in gases and meteorology. These studies led him to publish in 1808 asserting that matter is made up of atoms. All atoms of the same element are alike, having the same mass but differing from other elements. Elements combine in simple whole number proportions in making compounds.

Amedeo Avogadro 1776-1856   Avogadro was an Italian Count. He became a chemist and professor of physics. In 1811 Avogadro published what became known as Avogadro's hypothesis which claimed that at the same temperature and pressure, equal volumes of gas have equal numbers of molecules. European scientists did not accept this theory until four years after his death, when Avogadro's student, Stanislao Cannizzaro, convinced them of its validity. It is now known as Avogadro's Law.

Jons Jakob Berzelius 1779-1848   Berzelius was trained as a physician in his native Sweden. He became a professor of chemistry and secretary of the Stockholm Academy of Sciences. A prolific scientist, he accurately determined atomic masses, determined molecular structures, and investigated catalysts. Berzelius introduced the present system of chemical symbols of elements and formulas for compounds.

Dmitri Mendeleev 1834-1907   Mendeleev was the youngest child in a very large family in Siberia. He studied, and later taught, chemistry at the University of St. Petersburg in Russia. In 1869 he published his Periodic Table where the elements were arranged in families and in order of their atomic masses. His table contained some gaps for elements not yet discovered. Mendeleev predicted not only their existence but their properties as well.

Svante Arrhenius 1859-1927   Arrhenius was a Swedish chemist who began his PH.D. program in 1880. His doctoral thesis was on the electrical conduction of solutions. From his experimentation he deduced that ions were responsible and developed the theory of electrolytes and ionic bonding. His professors were skeptical and only reluctantly granted him his degree in 1884. Undeterred, Arrhenius continued to build support for his theory for which he won the Nobel Prize 19 years later.

Henrie Louis LeChatelier 1850-1936   LeChatelier trained as a mining engineer, and later became a professor of chemistry at the University of Paris. In 1888 he studied the effect of temperature and pressure on equilibrium from which he was able to maximize production in a reaction. He found that if changes are made to a system at equilibrium, the reaction tends to shift to attempt to restore original conditions.

Marie Curie 1867-1934 Curie was born and raised in Poland. She investigated the newly discovered radioactivity with her husband Pierre in Paris. They discovered the elements radium and polonium . Marie Curie worked on the chemical separation of radium from its compounds. She was awarded two Nobel Prizes.

Joseph John Thomson 1856-1940   Thomson was an English physicist at Cambridge University. He did extensive research with a cathode ray tube which produced a stream of negative particles. In 1897 Thomson called these particles electrons and postulated they were part of all atoms in equal number to their protons. His model had the electrons embedded among the positive charges as "plums in a pudding."

Ernest Rutherford 1971-1937   Rutherford was a native of New Zealand. He completed his education at Cambridge University. Rutherford was a professor of physics in Canada and England, working on radioactivity. He characterized alpha, beta, and gamma decay. By bombarding gold foil with positive alpha particles, he was able to deduce in 1911 that an atom has a very small, dense, positively charged nucleus with tiny electrons flying about the nucleus in orbits and that most of the atom was empty space. In 1919 Rutherford bombarded nitrogen gas with alpha particles and obtained oxygen atoms and protons. This was the first artificial transmutation.

Henry Moseley 1887-1915 Moseley was a child prodigy in England. He entered Eton College at age nine. Upon graduating from Oxford, Moseley went to work with Ernest Rutherford. Moseley studied the x-ray spectra emitted when metals were bombarded by streams of electrons. From this he deduced that elements have a unique charge on their nucleus, an atomic number. The periodic tables were then amended to show families organized by atomic number instead of mass. Tragically, Moseley's life and career ended too soon when he was killed in World War I.

Albert Einstein 1879-1955   Einstein rose from an uninspired student in Germany to become the greatest physicist of his time. He is best known for his theories of relativity. However, he won his Nobel Prize for work he published in 1905 on the photoelectric effect which showed that electrons accept certain quantities of energy. This led to the quantum mechanical model of the atom.

Niels Bohr 1885-196   Bohr was a Dane, but he studied atomic physics in England with Ernest Rutherford. By looking at the Hydrogen spectrum, he deduced that the hydrogen electron was in energy levels, but it could jump from level to level when excited with energy. The light energy the electron lost produced the spectrum. This led to the development of the solar system model of the atom.

Linus Pauling 1901-1994   Pauling was an American chemist who spent most of his career at Caltech. He gave us an understanding of the chemical bond. He developed the electronegativity scale to explain bonding. He described hydrogen bonding. He found the alpha-helix structure of proteins. He won a Nobel Prize in chemistry, but in 1962 he also won a Nobel Peace Prize for his work on stopping the testing of nuclear weapons.

Glenn T. Seaborg 1912-1999   Seaborg was born in Michigan and spent much of his career at the University of California.. He discovered many radioactive isotopes and was director of the Lawrence Berkeley Laboratory. With the use of the particle accelerators at that laboratory, Seaborg participated in the discovery of the heavy elements plutonium, americium, curium, berkelium, californium, einsteinium, fermium, mendelevium, and nobelium.

APPENDIX C

REFERENCE TABLES

Since reference tables are used so often by scientists and students of science, it is appropriate that high school chemistry students be familiar with them. The tables provided by the New York State Education Department for students taking the Regents Examination are presented here. A brief description of the type of information to be found and some test items based on each of the tables are also provided. However, students are urged to peruse the entire booklet of reference tables thoroughly while taking the examination. There is much additional information to be found in many of the tables. For example, tables may contain definitions, sample equations, the names and the formulas of ions and compounds, etc.

Some Regents test items contain phrases such as: "according to Reference Table G...," or "based on Reference Table N." Many other items require information contained in the tables. Students should be aware of this. Competency in information retrieval is one of the important objectives of the chemistry program.

TABLE A

Standard Temperature and Pressure

These are values used as standard condition for gases. They are useful in gas law problems. Always use the temperature in kelvins (see Table T) when used in gas law problems. STP means at conditions of standard temperature and pressure.

QUESTIONS

1. What is the volume of a gas at 150. kPa pressure if it was 2.50 L at standard pressure and constant temperature?

   _____

2. What is the pressure of 45.0 mL of a gas at 25°C if it had been 1.58 atm at 52°C in 37.0 mL?

   _____

3. What is the volume of oxygen gas at STP if it is 56.0 mL at 52°C and 1.58 atm?

_____

TABLE B

Physical Constants for Water

Heat of fusion is the amount of heat it takes to melt one gram of ice. The same amount of heat is released when it freezes (heat of solidification).

The _heat of vaporization_ is the amount of heat needed to boil one gram of water into gas at its boiling point. The same amount of heat is released as it condenses (heat of condensation).

The _specific heat capacity_ is the amount of heat it takes to raise one gram of liquid water one degree Celsius (see Table T). Notice that a change of temperature of one degree Celsius is the same temperature change as one Kelvin.

QUESTIONS

4. How much heat does it take to melt 5.0 g of ice?

_____

5. How many grams of water can 10. kJ boil into vapor?

_____

6. How much heat does it take to heat 5.0 g of water from 15.0°C to 28.0°C?

_____

TABLE C

The prefixes and symbols are listed for exponential factors indicating the size of the quantity measured.

7. 2×10^{-6} Joules is how many microjoules?

_____

8. 4.18 kL is how many liters?

_____

9. How many centimeters is 45 km?

_____

TABLE D
Selected Units

This table gives the unit and symbol for important quantities used in chemistry. The units listed are mostly standard international units.

QUESTIONS

10. What is the standard international unit for heat?

_____

11. L is the symbol used for what unit?

_____

12. What is the name and symbol for the standard international unit for pressure?

_____

12a. What is the symbol used for atomic mass units?

_____

TABLE E
Selected Polyatomic Ions

Table E has many applications in naming compounds and writing their formulas. Since the charge on each ion is presented, the table is also helpful in determining oxidation numbers. Notice that the convention in the notation for ionic charge is as follows:

1. When the charge is 1+ or 1–, only the sign is given, not the numeral.

2. For charges greater than one, the numeral is written first, then the charge. For example, 2+ or 3–.

Oxidation numbers are generally designated sign first, then numeral.
For example, +1, −1, +2, −2, etc. (See Periodic Table of the Elements.)

When naming acids of polyatomic ions the rules are:

1. If the name of the polyatomic ion ends in –*ite*,
   the name of the acid associated with it ends in –*ous*.
   For example, nitr*ous* acid produces nitr*ite* ions.

2. If the name of the ion ends in –*ate*,
   the name of the acid ends in –*ic*.
   Nitr*ic* acid produces nitr*ate* ions.

QUESTIONS

13. The correct formula for the thiosulfate ion is

    (1) SO_3^{2-} (2) SO_4^{2-} (3) SCN^- (4) $S_2O_3^{2-}$

14. Which formula correctly represents mercury (I) chloride?

    (1) Hg_2Cl (2) $HgCl_2$ (3) Hg_2Cl_2 (4) Hg_2Cl_4

15. What is the name of the calcium salt of sulfuric acid?

    (1) calcium thisulfate (2) calcium sulfate (3) calcium sulfide (4) calcium sulfite

16. What is the correct formula for sodium thiosulfate?

    (1) $Na_2S_2O_4$ (2) Na_2SO_3 (3) Na_2SO_4 (4) $Na_2S_2O_3$

TABLE F
Solubility Guidelines

Solubility rules allow the prediction of precipitation or dissolving upon mixing of ionic solution at moderate concentrations (about 0.1 M to 1.0 M). If it dissolves it will not precipitate. Match a positive ion with a negative ion to use the table.

QUESTIONS

17. Which of these compounds will readily dissolve in water?

    (1) Na_3PO_4 (2) $CaCO_3$ (3) $PbCl_2$ (4) $Fe(OH)_2$

18. What will most likely happen upon mixing aqueous solutions of $(NH_4)_2S$ and $CaBr_2$?

    (1) no precipitate occurs   (2) NH_4Br precipitates
    (3) CaS precipitates   (4) $CaBr_2$ precipitates

19. Which will not readily dissolve in water?

    (1) $AgHCO_3$ (2) Ag_2SO_4 (3) $AgNO_3$ (4) LiOH

TABLE G
Solubility curves

Each of the curves in this table represents the concentration, in grams of solute per 100 grams water, of *saturated* solutions at various temperatures.

Some helpful hints:

1. Be certain to understand the definition of the y axis.

2. Questions often present names, rather than formulas, of compounds. If this represents a problem, be sure to take advantage of the information found in tables E, F, and S.

3. Note that most solids are more soluble at higher temperatures.

4. The gases in the table, HCl, NH_3 (ammonia), and SO_2, show a decrease in solubility with an increase in temperature. This behavior is typical of gases.

QUESTIONS

20. According to Reference Table G, approximately how many grams of $KClO_3$ are needed to saturate 100. grams of H_2O at 40.°C?

    (1) 6   (2) 16   (3) 38   (4) 47

21. How many grams of KNO_3 are needed to saturate 50. grams of water at 70.°C?

    (1) 30. g   (2) 65 g   (3) 130.g   (4) 160.g

22. Which quantity of salt will form a saturated solution of 100. grams of water at 45°C?

    (1) 30.g of KCl   (2) 35 g of NH_4Cl (3) 60.g of KNO_3 (4) 110.g of $NaNO_3$

23. A solution containing 90. grams of KNO_3 in 100. grams of H_2O at 50.°C is considered to be

    (1) dilute and unsaturated   (2) dilute and supersaturated
    (3) concentrated and unsaturated   (4) concentrated and supersaturated

24. A solution contains 14 grams of KCl in 100. grams of water at 40.°C. What is the minimum amount of KCl that must be added to make this a saturated solution?

    (1) 14 g   (2) 19 g   (3) 25 g   (4) 44 g

25. According to Reference Table G, how does a decrease in temperature from 40.°C to 20.°C affect the solubility of NH_3 and KCl?

    (1) The solubility of NH_3 decreases, and the solubility of KCl decreases.
    (2) The solubility of NH_3 decreases and the solubility of KCl increases.
    (3) The solubility of NH_3 increases and the solubility of KCl decreases.
    (4) The solubility of NH_3 increases, and solubility of KCl increases.

TABLE H
Vapor Pressure of Four Liquids

    The vapor pressure of liquids at a particular temperature is a measure of the strength of the intermolecular forces between their molecules. If the vapor pressure is high, the attraction between molecules must be weak. The molecules can escape the liquid easily to become a gas, and the vapor pressure is higher. Liquids boil when their vapor pressure is equal to (or greater than) the atmospheric pressure around the liquid.

QUESTIONS

26. What is the boiling point of ethanol at standard pressure?

_____

27. Which of the four liquids on Table H has the strongest intermolecular forces?

_____

28. At what temperature will propanone boil if the atmospheric pressure is 70. kPa?

_____

TABLE I
Heats of Reaction at 101.3 kPa and 298K

The values for the heats of reaction are for one mole of the compound reacting (or if only elements are reacting, for one mole of the compound being formed as a product) _if_ the equation is balanced for one mole of compound. However, if the equations is balanced with a coefficient of 2 in front of the compound, then the heat of reaction is for two moles of compound reacted or formed.

Helpful Hints:

1. Where ΔH is _negative_, the reaction is _exothermic_.

2. Where ΔH is _positive_, the reaction is _endothermic_.

3. Exothermic reactions add heat to the environment and, therefore, raise the temperature of the reaction medium. Dissolving NaOH in water causes the temperature of the water to rise.

4. Endothermic reactions remove heat from the environment. Dissolving NH_4Cl in water lowers the temperature of the water.

5. By multiplying an equation by a factor, one can determine the heat produced (or absorbed) by multiplying ΔH by the same factor. For example, multiplying the first equation by 2 gives $2CH_4(g) + 4O_2(g) \rightarrow 2CO_2(g) + 4H_2O(\ell)$ and $\Delta H = 2 (-890.4kJ)$ or -1780.0 kJ

QUESTIONS

29. The greatest amount of energy would be given up by the complete oxidation of

(1) $CH_4(g)$ (2) $C_3H_8(g)$ (3) CH_3OH (4) $C_6H_{12}O_6(s)$

30. When $KNO_3(s)$ is dissolved in water, the temperature of the water will

(1) increase   (2) decrease   (3) remain the same

31. How much heat is required to form one mole of NO(g) out of its elements?

_____

31a. How much heat is released when 1 mole of Al(s) reacts with sufficient O_2 (g) to produce $Al_2O_3(s)$?

_____

TABLE J
Activity Series

An element on the list is more reactive than any element below it. The higher neutral metal elements will be oxidized by metal ions below them. The higher neutral non-metals will be reduced by the non-metal ions below them. In a single replacement reaction, a higher element will replace a lower element in a compound.

QUESTIONS

32. Will Ba react with Mn^{2+}?

_____

33. Will Na^+ react with Cr?

_____

34. What are the products of the reaction $Mg + Co(NO_3)_2$?

_____

TABLE K AND TABLE L
Common Acids and Common Bases

Acids and bases are electrolytes. The positive ion in acids is H^+. The negative ion in bases is OH^-. Notice that $H_2O + CO_2$ is H_2CO_3 and $H_2O + NH_3$ is NH_4OH. The bottom two acids (carbonic acid and ethanoic acid) and the bottom base (ammonia) are weak. The others listed are strong.

TABLE M
Common Acid-Base Indicators

Acid-base indicators are organic dyes that are a type of acid or base themselves. They occur in two forms - one form (and color) when in an acid and another form (and color) when in a base. The first color mentioned for each indicator corresponds to the lower pH of its range and any lower pH. That is its acid indicating color. The second color mentioned corresponds to the indicator's color at the higher pH given and any pH higher. This is the indicator's base range. At a pH inside an indicator's color range, a mixture of the two forms and colors result. When titrating an acid with a base, one tries to use an indicator that changes color at the pH of the equivalency point.

QUESTIONS

35. What is the color of phenolphthalein in a solution whose pH is 11?

_____

36. Thymol blue would be yellow in what pH solution?

_____

37. Which indicator would be best for an acid-base titration whose equivalency point is at a pH of 6.5?

(1) methyl orange   (2) bromthymol blue (3) phenolphthalein   (4) thymol blue

TABLE N
Selected Radioisotopes

Table N is designed to be used in calculating rates, times, and/or masses involved during the decay of samples of the isotopes specified in a variety of conditions. Nuclide is a term used to refer to any isotope of an element. The particular isotope is indicated by the mass number, which is written as a superscript to the left of the symbol. ^{14}C is read "carbon fourteen," and refers to the isotope of carbon containing 8 neutrons.

Remember: mass number = protons + neutrons.

All isotopes of carbon contain 6 protons. Thus, $14 - 6 = 8$ neutrons.
The key to utilizing this chart is the statement:

> At the conclusion of each half-life, the mass of the sample is half of the mass it had at the beginning of that half-life.

(See Chapter 10)

QUESTIONS

38. What is the number of h required for potassium-42 to undergo 3 half-life periods?

(1) 6.2 h (2) 12.4 h (3) 24.7 h (4) 37.1 h

39. How many grams of a 32-gram sample of ^{32}P will remain after 71.2 d?

(1) 1 (2) 2 (3) 8 (4) 4

TABLE O
Symbols Used in Nuclear Chemistry

The notation of particles gives the mass number (protons + neutrons) as the upper number and charge as the lower number. This chart together with Table N can be used to write nuclear equations and determine products. (See Chapter 10)

QUESTIONS

40. Which two particles have approximately the same mass?

(1) neutron and electron (2) neutron and deuteron
(3) proton and neutron (4) proton and electron

41. Which nuclear emission moving though an electric field would be deflected toward the positive electrode?

(1) alpha particle (2) beta particle (3) gamma radiation (4) proton

42. Which particle is electrically neutral?

(1) proton (2) positron (3) neutron (4) electron

TABLE P
Organic Prefixes

The prefixes are the organic chemistry way of counting carbon atoms in a chain or branch. They can be used together with Table Q to name all chain hydrocarbons up to 10 carbon atoms long. Table P can also be used with Table R to name many other organic compounds. The prefix for the number of carbon atoms is the beginning part of all organic names of compounds from Table Q and R.

TABLE Q
Homologous Series of Hydrocarbons

Hydrocarbons have only the elements hydrogen and carbon. The alkanes have only single bonds, and their names all end in -ane. The alkenes have a double bond in the chain, and their names all end in -ene. The alkynes have a triple bond in the chain, and their names all end in -yne. All hydrogens bond once, and all carbons bond four times. In graphic structural formulas, the bonds are represented by lines. Single bonds have one line, double bonds have two lines, and triple bonds have three lines. If isomers are possible with a different location of the double or triple bond, the carbons must be numbered. (See Chapter 7)

QUESTIONS

43-45   Name these hydrocarbons:

```
         H  H  H
         |  |  |
43.   H- C- C- C- H
         |  |  |
         H  H  H
```

_____

```
         H  H  H  H
         |  |  |  |
44.   H- C- C= C- C- H
         |        |
         H        H
```

_____

```
         H  H  H        H  H
         |  |  |        |  |
      H- C- C- C- C≡ C- C- C- H
         |  |  |        |  |
         H  H  H        H  H
45.
```

_____

46. Write the molecular formula for 2-octene.

TABLE R
Organic Functional Groups

The difference in chemical properties between organic compounds is mostly due to their functional group. The rest of all organic molecules is just hydrocarbon chains. In the names, the beginning part is the hydrocarbon part, and the ending is unique for the functional group. All alcohols end in -ol. All aldehydes end in -al, etc. This is not the case, however, for esters and ethers. With the functional group in the middle, these compounds use both hydrocarbon ends in the name. (See Chapter 7)

QUESTIONS

47-50 Name these compounds:

47.
```
    H H H   O
    | | |  //
H-C-C-C-C-H
    | | |
    H H H
```
_____

48.
```
    H OH H H H
    | |  | | |
H-C- C -C-C-C-H
    | |  | | |
    H H  H H H
```
_____

49.
```
    H H Br Br H
    | |  |  | |
H-C-C- C - C -C-H
    | |  |  | |
    H H  H  H H
```
_____

50.
```
    H H H
    | | |
H-C-C-N-H
    | |
    H H
```
_____

51. Give the structural formula for ethyl methanoate.

52. What class of compound is 3-octanone?

TABLE S
Properties of Selected Elements

This table supplies the names and symbols of all the important elements and several of their physical properties. Often the physical properties can help predict chemical properties as well. For example, if the ionization energy and electronegativity are high, the element attracts electrons strongly. It is likely to form negative ions and be a non-metal. If the ionization energy and electronegativity are low, the element will form a positive ion and be a metal. If the melting and boiling points are high, they must have strong intermolecular forces. The densities of the elements are given at STP so if it is very low, the element is a gas. If the melting point is above room temperature, the element is a solid. If the boiling point is below room temperature, the element is a gas. If room temperature is between the melting and boiling points, the element is a liquid.

QUESTIONS

53. What element has the highest attraction for electrons in a chemical bond?

_____

54. Which element has the biggest atoms?

(1) S   (2) Cl   (3) K   (4) Ca

55. What is the minimum energy needed to remove an electron from a neutral potassium atom?

_____

56. Name two common elements which are liquid at room temperature.

_____   _____

57. How many meters is the radius of a hydrogen atom?

_____

TABLE T
Important Formulas and Equations

These formulas allow one to find the relationship needed to do many calculations without committing the formulas to memory. However, unless a student has practiced using them during the year, the formulas do not always make sense if confronting them for the first time at a test. The combined gas laws can be used to solve volume-temperature, volume-pressure, and temperature-pressure problems also. Just divide out the term that does not change. The temperature must be converted to Kelvins if using the gas laws. The specific heat capacity, heat of vaporization, and heat of fusion for water are given on Table B for doing heat calculations. Percent error is commonly done with laboratory results to find the accuracy of an experiment. The density equation shows how the last column of Table S is determined.

QUESTIONS

58. A student's experiment shows the volume of 1.00 mol of gas at STP to be 26.3 L. The accepted value is 22.4 L. What is the student's percent error?

_____

59. How many grams of salt must be in 500. g of solution to have a concentration of 100. ppm?

_____

60. What is the pressure of nitrogen gas at 50.°C if the pressure is 800. kPa at 100.°C with constant volume?

_____

61. How much heat is needed to melt 100. g of water ice?

_____

62. What volume of 3.5 **M** NaOH(aq) is needed to titrate 15 mL of 4.5 **M** HCl(aq) to the equivalency point?

_____

THE UNIVERSITY OF THE STATE OF NEW YORK • THE STATE EDUCATION DEPARTMENT • ALBANY, NY 12234

C
Reference Tables for Physical Setting/CHEMISTRY
2011 Edition

Table A
Standard Temperature and Pressure

Name	Value	Unit
Standard Pressure	101.3 kPa 1 atm	kilopascal atmosphere
Standard Temperature	273 K 0°C	kelvin degree Celsius

Table B
Physical Constants for Water

Heat of Fusion	334 J/g
Heat of Vaporization	2260 J/g
Specific Heat Capacity of $H_2O(\ell)$	4.18 J/g•K

Table C
Selected Prefixes

Factor	Prefix	Symbol
10^3	kilo-	k
10^{-1}	deci-	d
10^{-2}	centi-	c
10^{-3}	milli-	m
10^{-6}	micro-	μ
10^{-9}	nano-	n
10^{-12}	pico-	p

Table D
Selected Units

Symbol	Name	Quantity
m	meter	length
g	gram	mass
Pa	pascal	pressure
K	kelvin	temperature
mol	mole	amount of substance
J	joule	energy, work, quantity of heat
s	second	time
min	minute	time
h	hour	time
d	day	time
y	year	time
L	liter	volume
ppm	parts per million	concentration
M	molarity	solution concentration
u	atomic mass unit	atomic mass

Table E
Selected Polyatomic Ions

Formula	Name	Formula	Name
H_3O^+	hydronium	CrO_4^{2-}	chromate
Hg_2^{2+}	mercury(I)	$Cr_2O_7^{2-}$	dichromate
NH_4^+	ammonium	MnO_4^-	permanganate
$C_2H_3O_2^-$ CH_3COO^- } acetate		NO_2^-	nitrite
		NO_3^-	nitrate
CN^-	cyanide	O_2^{2-}	peroxide
CO_3^{2-}	carbonate	OH^-	hydroxide
HCO_3^-	hydrogen carbonate	PO_4^{3-}	phosphate
$C_2O_4^{2-}$	oxalate	SCN^-	thiocyanate
ClO^-	hypochlorite	SO_3^{2-}	sulfite
ClO_2^-	chlorite	SO_4^{2-}	sulfate
ClO_3^-	chlorate	HSO_4^-	hydrogen sulfate
ClO_4^-	perchlorate	$S_2O_3^{2-}$	thiosulfate

Table F
Solubility Guidelines for Aqueous Solutions

Ions That Form *Soluble* Compounds	Exceptions	Ions That Form *Insoluble* Compounds*	Exceptions
Group 1 ions (Li^+, Na^+, etc.)		carbonate (CO_3^{2-})	when combined with Group 1 ions or ammonium (NH_4^+)
ammonium (NH_4^+)		chromate (CrO_4^{2-})	when combined with Group 1 ions, Ca^{2+}, Mg^{2+}, or ammonium (NH_4^+)
nitrate (NO_3^-)			
acetate ($C_2H_3O_2^-$ or CH_3COO^-)		phosphate (PO_4^{3-})	when combined with Group 1 ions or ammonium (NH_4^+)
hydrogen carbonate (HCO_3^-)		sulfide (S^{2-})	when combined with Group 1 ions or ammonium (NH_4^+)
chlorate (ClO_3^-)		hydroxide (OH^-)	when combined with Group 1 ions, Ca^{2+}, Ba^{2+}, Sr^{2+}, or ammonium (NH_4^+)
halides (Cl^-, Br^-, I^-)	when combined with Ag^+, Pb^{2+}, or Hg_2^{2+}		
sulfates (SO_4^{2-})	when combined with Ag^+, Ca^{2+}, Sr^{2+}, Ba^{2+}, or Pb^{2+}		

*compounds having very low solubility in H_2O

REFERENCE TABLES

Table G
Solubility Curves at Standard Pressure

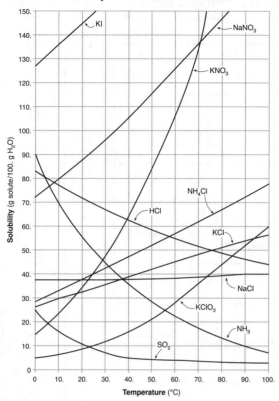

Table H
Vapor Pressure of Four Liquids

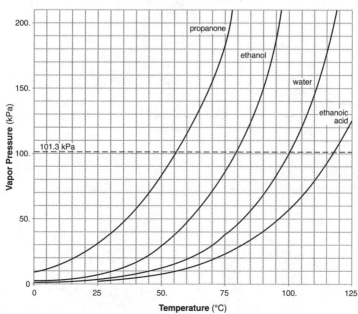

Table I
Heats of Reaction at 101.3 kPa and 298 K

Reaction	ΔH (kJ)*
$CH_4(g) + 2O_2(g) \longrightarrow CO_2(g) + 2H_2O(\ell)$	−890.4
$C_3H_8(g) + 5O_2(g) \longrightarrow 3CO_2(g) + 4H_2O(\ell)$	−2219.2
$2C_8H_{18}(\ell) + 25O_2(g) \longrightarrow 16CO_2(g) + 18H_2O(\ell)$	−10943
$2CH_3OH(\ell) + 3O_2(g) \longrightarrow 2CO_2(g) + 4H_2O(\ell)$	−1452
$C_2H_5OH(\ell) + 3O_2(g) \longrightarrow 2CO_2(g) + 3H_2O(\ell)$	−1367
$C_6H_{12}O_6(s) + 6O_2(g) \longrightarrow 6CO_2(g) + 6H_2O(\ell)$	−2804
$2CO(g) + O_2(g) \longrightarrow 2CO_2(g)$	−566.0
$C(s) + O_2(g) \longrightarrow CO_2(g)$	−393.5
$4Al(s) + 3O_2(g) \longrightarrow 2Al_2O_3(s)$	−3351
$N_2(g) + O_2(g) \longrightarrow 2NO(g)$	+182.6
$N_2(g) + 2O_2(g) \longrightarrow 2NO_2(g)$	+66.4
$2H_2(g) + O_2(g) \longrightarrow 2H_2O(g)$	−483.6
$2H_2(g) + O_2(g) \longrightarrow 2H_2O(\ell)$	−571.6
$N_2(g) + 3H_2(g) \longrightarrow 2NH_3(g)$	−91.8
$2C(s) + 3H_2(g) \longrightarrow C_2H_6(g)$	−84.0
$2C(s) + 2H_2(g) \longrightarrow C_2H_4(g)$	+52.4
$2C(s) + H_2(g) \longrightarrow C_2H_2(g)$	+227.4
$H_2(g) + I_2(g) \longrightarrow 2HI(g)$	+53.0
$KNO_3(s) \xrightarrow{H_2O} K^+(aq) + NO_3^-(aq)$	+34.89
$NaOH(s) \xrightarrow{H_2O} Na^+(aq) + OH^-(aq)$	−44.51
$NH_4Cl(s) \xrightarrow{H_2O} NH_4^+(aq) + Cl^-(aq)$	+14.78
$NH_4NO_3(s) \xrightarrow{H_2O} NH_4^+(aq) + NO_3^-(aq)$	+25.69
$NaCl(s) \xrightarrow{H_2O} Na^+(aq) + Cl^-(aq)$	+3.88
$LiBr(s) \xrightarrow{H_2O} Li^+(aq) + Br^-(aq)$	−48.83
$H^+(aq) + OH^-(aq) \longrightarrow H_2O(\ell)$	−55.8

*The ΔH values are based on molar quantities represented in the equations. A minus sign indicates an exothermic reaction.

Table J
Activity Series**

Most Active	Metals	Nonmetals	Most Active
	Li	F_2	
	Rb	Cl_2	
	K	Br_2	
	Cs	I_2	
	Ba		
	Sr		
	Ca		
	Na		
	Mg		
	Al		
	Ti		
	Mn		
	Zn		
	Cr		
	Fe		
	Co		
	Ni		
	Sn		
	Pb		
	H_2		
	Cu		
	Ag		
Least Active	Au		Least Active

**Activity Series is based on the hydrogen standard. H_2 is *not* a metal.

Table K
Common Acids

Formula	Name
HCl(aq)	hydrochloric acid
HNO_2(aq)	nitrous acid
HNO_3(aq)	nitric acid
H_2SO_3(aq)	sulfurous acid
H_2SO_4(aq)	sulfuric acid
H_3PO_4(aq)	phosphoric acid
H_2CO_3(aq) or CO_2(aq)	carbonic acid
CH_3COOH(aq) or $HC_2H_3O_2$(aq)	ethanoic acid (acetic acid)

Table L
Common Bases

Formula	Name
NaOH(aq)	sodium hydroxide
KOH(aq)	potassium hydroxide
$Ca(OH)_2$(aq)	calcium hydroxide
NH_3(aq)	aqueous ammonia

Table M
Common Acid–Base Indicators

Indicator	Approximate pH Range for Color Change	Color Change
methyl orange	3.1–4.4	red to yellow
bromthymol blue	6.0–7.6	yellow to blue
phenolphthalein	8–9	colorless to pink
litmus	4.5–8.3	red to blue
bromcresol green	3.8–5.4	yellow to blue
thymol blue	8.0–9.6	yellow to blue

Source: *The Merck Index*, 14th ed., 2006, Merck Publishing Group

Table N
Selected Radioisotopes

Nuclide	Half-Life	Decay Mode	Nuclide Name
^{198}Au	2.695 d	β^-	gold-198
^{14}C	5715 y	β^-	carbon-14
^{37}Ca	182 ms	β^+	calcium-37
^{60}Co	5.271 y	β^-	cobalt-60
^{137}Cs	30.2 y	β^-	cesium-137
^{53}Fe	8.51 min	β^+	iron-53
^{220}Fr	27.4 s	α	francium-220
^{3}H	12.31 y	β^-	hydrogen-3
^{131}I	8.021 d	β^-	iodine-131
^{37}K	1.23 s	β^+	potassium-37
^{42}K	12.36 h	β^-	potassium-42
^{85}Kr	10.73 y	β^-	krypton-85
^{16}N	7.13 s	β^-	nitrogen-16
^{19}Ne	17.22 s	β^+	neon-19
^{32}P	14.28 d	β^-	phosphorus-32
^{239}Pu	2.410×10^4 y	α	plutonium-239
^{226}Ra	1599 y	α	radium-226
^{222}Rn	3.823 d	α	radon-222
^{90}Sr	29.1 y	β^-	strontium-90
^{99}Tc	2.13×10^5 y	β^-	technetium-99
^{232}Th	1.40×10^{10} y	α	thorium-232
^{233}U	1.592×10^5 y	α	uranium-233
^{235}U	7.04×10^8 y	α	uranium-235
^{238}U	4.47×10^9 y	α	uranium-238

Source: *CRC Handbook of Chemistry and Physics*, 91st ed., 2010–2011, CRC Press

Table O
Symbols Used in Nuclear Chemistry

Name	Notation	Symbol
alpha particle	^4_2He or $^4_2\alpha$	α
beta particle	$^0_{-1}e$ or $^0_{-1}\beta$	β^-
gamma radiation	$^0_0\gamma$	γ
neutron	^1_0n	n
proton	^1_1H or ^1_1p	p
positron	$^0_{+1}e$ or $^0_{+1}\beta$	β^+

Table P
Organic Prefixes

Prefix	Number of Carbon Atoms
meth-	1
eth-	2
prop-	3
but-	4
pent-	5
hex-	6
hept-	7
oct-	8
non-	9
dec-	10

Table Q
Homologous Series of Hydrocarbons

Name	General Formula	Examples	
		Name	Structural Formula
alkanes	C_nH_{2n+2}	ethane	H H | | H—C—C—H | | H H
alkenes	C_nH_{2n}	ethene	H H \ / C=C / \ H H
alkynes	C_nH_{2n-2}	ethyne	H—C≡C—H

Note: n = number of carbon atoms

Table R
Organic Functional Groups

Class of Compound	Functional Group	General Formula	Example
halide (halocarbon)	$-F$ (fluoro-) $-Cl$ (chloro-) $-Br$ (bromo-) $-I$ (iodo-)	$R-X$ (X represents any halogen)	$CH_3CHClCH_3$ 2-chloropropane
alcohol	$-OH$	$R-OH$	$CH_3CH_2CH_2OH$ 1-propanol
ether	$-O-$	$R-O-R'$	$CH_3OCH_2CH_3$ methyl ethyl ether
aldehyde	$-\overset{\overset{O}{\|}}{C}-H$	$R-\overset{\overset{O}{\|}}{C}-H$	$CH_3CH_2\overset{\overset{O}{\|}}{C}-H$ propanal
ketone	$-\overset{\overset{O}{\|}}{C}-$	$R-\overset{\overset{O}{\|}}{C}-R'$	$CH_3\overset{\overset{O}{\|}}{C}CH_2CH_2CH_3$ 2-pentanone
organic acid	$-\overset{\overset{O}{\|}}{C}-OH$	$R-\overset{\overset{O}{\|}}{C}-OH$	$CH_3CH_2\overset{\overset{O}{\|}}{C}-OH$ propanoic acid
ester	$-\overset{\overset{O}{\|}}{C}-O-$	$R-\overset{\overset{O}{\|}}{C}-O-R'$	$CH_3CH_2\overset{\overset{O}{\|}}{C}OCH_3$ methyl propanoate
amine	$-\overset{\overset{\|}{N}}{\|}-$	$R-\overset{\overset{R'}{\|}}{N}-R''$	$CH_3CH_2CH_2NH_2$ 1-propanamine
amide	$-\overset{\overset{O}{\|}}{C}-NH$	$R-\overset{\overset{O}{\|}}{C}-NH$ (R')	$CH_3CH_2\overset{\overset{O}{\|}}{C}-NH_2$ propanamide

Note: R represents a bonded atom or group of atoms.

Periodic Table

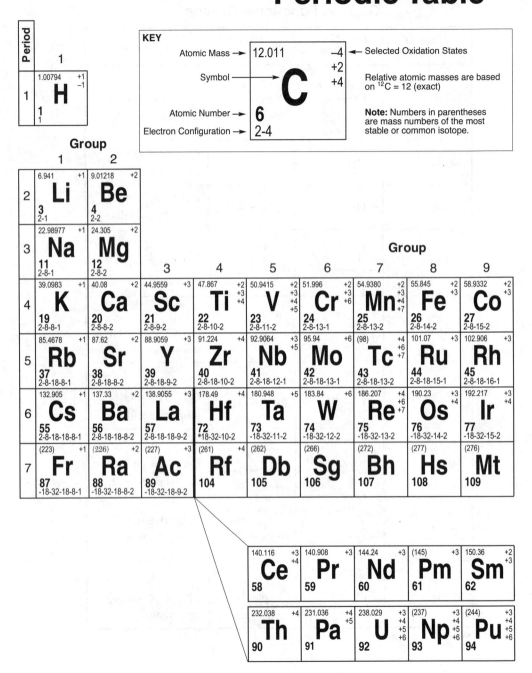

KEY

Atomic Mass → 12.011   −4 ← Selected Oxidation States
    +2
Symbol → **C**   +4   Relative atomic masses are based on ^{12}C = 12 (exact)

Atomic Number → **6**

Electron Configuration → 2-4   **Note:** Numbers in parentheses are mass numbers of the most stable or common isotope.

*denotes the presence of (2-8-) for elements 72 and above

**The systematic names and symbols for elements of atomic numbers 113 and above will be used until the approval of trivial names by IUPAC.

Source: *CRC Handbook of Chemistry and Physics*, 91st ed., 2010–2011, CRC Press

of the Elements

						18
						4.00260 0 **He** 2 2

Group

	13	14	15	16	17	18
	10.81 +3 **B** 5 2-3	12.011 −4 / +2 / +4 **C** 6 2-4	14.0067 −3/−2/−1/+1/+2/+3/+4/+5 **N** 7 2-5	15.9994 −2 **O** 8 2-6	18.9984 −1 **F** 9 2-7	20.180 0 **Ne** 10 2-8
	26.98154 +3 **Al** 13 2-8-3	28.0855 −4/+2/+4 **Si** 14 2-8-4	30.97376 −3/+3/+5 **P** 15 2-8-5	32.065 −2/+4/+6 **S** 16 2-8-6	35.453 −1/+1/+5/+7 **Cl** 17 2-8-7	39.948 0 **Ar** 18 2-8-8

10	11	12	13	14	15	16	17	18
58.693 +2/+3 **Ni** 28 2-8-16-2	63.546 +1/+2 **Cu** 29 2-8-18-1	65.409 +2 **Zn** 30 2-8-18-2	69.723 +3 **Ga** 31 2-8-18-3	72.64 −4/+4 **Ge** 32 2-8-18-4	74.9216 −3/+3/+5 **As** 33 2-8-18-5	78.96 −2/+4/+6 **Se** 34 2-8-18-6	79.904 −1/+1/+5 **Br** 35 2-8-18-7	83.798 0/+2 **Kr** 36 2-8-18-8
106.42 +2/+4 **Pd** 46 2-8-18-18	107.868 +1 **Ag** 47 2-8-18-18-1	112.41 +2 **Cd** 48 2-8-18-18-2	114.818 +3 **In** 49 2-8-18-18-3	118.71 +2/+4 **Sn** 50 2-8-18-18-4	121.760 −3/+3/+5 **Sb** 51 2-8-18-18-5	127.60 −2/+4/+6 **Te** 52 2-8-18-18-6	126.904 −1/+1/+5/+7 **I** 53 2-8-18-18-7	131.29 0/+4/+6 **Xe** 54 2-8-18-18-8
195.08 +2/+4 **Pt** 78 -18-32-17-1	196.967 +1/+3 **Au** 79 -18-32-18-1	200.59 +1/+2 **Hg** 80 -18-32-18-2	204.383 +1/+3 **Tl** 81 -18-32-18-3	207.2 +2/+4 **Pb** 82 -18-32-18-4	208.980 +3/+5 **Bi** 83 -18-32-18-5	(209) +2/+4 **Po** 84 -18-32-18-6	(210) **At** 85 -18-32-18-7	(222) 0 **Rn** 86 -18-32-18-8
(281) **Ds** 110	(280) **Rg** 111	(285) **Cn** 112	(284) **Uut** 113**	(289) **Uuq** 114	(288) **Uup** 115	(292) **Uuh** 116	(?) **Uus** 117	(294) **Uuo** 118

151.964 +2/+3 **Eu** 63	157.25 +3 **Gd** 64	158.925 +3 **Tb** 65	162.500 +3 **Dy** 66	164.930 +3 **Ho** 67	167.259 +3 **Er** 68	168.934 +3 **Tm** 69	173.04 +2/+3 **Yb** 70	174.9668 +3 **Lu** 71
(243) +3/+4/+5/+6 **Am** 95	(247) +3 **Cm** 96	(247) +3/+4 **Bk** 97	(251) +3 **Cf** 98	(252) +3 **Es** 99	(257) +3 **Fm** 100	(258) +2/+3 **Md** 101	(259) +2/+3 **No** 102	(262) +3 **Lr** 103

Table S
Properties of Selected Elements

Atomic Number	Symbol	Name	First Ionization Energy (kJ/mol)	Electro-negativity	Melting Point (K)	Boiling* Point (K)	Density** (g/cm³)	Atomic Radius (pm)
1	H	hydrogen	1312	2.2	14	20.	0.000082	32
2	He	helium	2372	—	—	4	0.000164	37
3	Li	lithium	520.	1.0	454	1615	0.534	130.
4	Be	beryllium	900.	1.6	1560.	2744	1.85	99
5	B	boron	801	2.0	2348	4273	2.34	84
6	C	carbon	1086	2.6	—	—	—	75
7	N	nitrogen	1402	3.0	63	77	0.001145	71
8	O	oxygen	1314	3.4	54	90.	0.001308	64
9	F	fluorine	1681	4.0	53	85	0.001553	60.
10	Ne	neon	2081	—	24	27	0.000825	62
11	Na	sodium	496	0.9	371	1156	0.97	160.
12	Mg	magnesium	738	1.3	923	1363	1.74	140.
13	Al	aluminum	578	1.6	933	2792	2.70	124
14	Si	silicon	787	1.9	1687	3538	2.3296	114
15	P	phosphorus (white)	1012	2.2	317	554	1.823	109
16	S	sulfur (monoclinic)	1000.	2.6	388	718	2.00	104
17	Cl	chlorine	1251	3.2	172	239	0.002898	100.
18	Ar	argon	1521	—	84	87	0.001633	101
19	K	potassium	419	0.8	337	1032	0.89	200.
20	Ca	calcium	590.	1.0	1115	1757	1.54	174
21	Sc	scandium	633	1.4	1814	3109	2.99	159
22	Ti	titanium	659	1.5	1941	3560.	4.506	148
23	V	vanadium	651	1.6	2183	3680.	6.0	144
24	Cr	chromium	653	1.7	2180.	2944	7.15	130.
25	Mn	manganese	717	1.6	1519	2334	7.3	129
26	Fe	iron	762	1.8	1811	3134	7.87	124
27	Co	cobalt	760.	1.9	1768	3200.	8.86	118
28	Ni	nickel	737	1.9	1728	3186	8.90	117
29	Cu	copper	745	1.9	1358	2835	8.96	122
30	Zn	zinc	906	1.7	693	1180.	7.134	120.
31	Ga	gallium	579	1.8	303	2477	5.91	123
32	Ge	germanium	762	2.0	1211	3106	5.3234	120.
33	As	arsenic (gray)	944	2.2	1090.	—	5.75	120.
34	Se	selenium (gray)	941	2.6	494	958	4.809	118
35	Br	bromine	1140.	3.0	266	332	3.1028	117
36	Kr	krypton	1351	—	116	120.	0.003425	116
37	Rb	rubidium	403	0.8	312	961	1.53	215
38	Sr	strontium	549	1.0	1050.	1655	2.64	190.
39	Y	yttrium	600.	1.2	1795	3618	4.47	176
40	Zr	zirconium	640.	1.3	2128	4682	6.52	164

REFERENCE TABLES

Atomic Number	Symbol	Name	First Ionization Energy (kJ/mol)	Electro-negativity	Melting Point (K)	Boiling* Point (K)	Density** (g/cm³)	Atomic Radius (pm)
41	Nb	niobium	652	1.6	2750.	5017	8.57	156
42	Mo	molybdenum	684	2.2	2896	4912	10.2	146
43	Tc	technetium	702	2.1	2430.	4538	11	138
44	Ru	ruthenium	710.	2.2	2606	4423	12.1	136
45	Rh	rhodium	720.	2.3	2237	3968	12.4	134
46	Pd	palladium	804	2.2	1828	3236	12.0	130.
47	Ag	silver	731	1.9	1235	2435	10.5	136
48	Cd	cadmium	868	1.7	594	1040.	8.69	140.
49	In	indium	558	1.8	430.	2345	7.31	142
50	Sn	tin (white)	709	2.0	505	2875	7.287	140.
51	Sb	antimony (gray)	831	2.1	904	1860.	6.68	140.
52	Te	tellurium	869	2.1	723	1261	6.232	137
53	I	iodine	1008	2.7	387	457	4.933	136
54	Xe	xenon	1170.	2.6	161	165	0.005366	136
55	Cs	cesium	376	0.8	302	944	1.873	238
56	Ba	barium	503	0.9	1000.	2170.	3.62	206
57	La	lanthanum	538	1.1	1193	3737	6.15	194
Elements 58–71 have been omitted.								
72	Hf	hafnium	659	1.3	2506	4876	13.3	164
73	Ta	tantalum	728	1.5	3290.	5731	16.4	158
74	W	tungsten	759	1.7	3695	5828	19.3	150.
75	Re	rhenium	756	1.9	3458	5869	20.8	141
76	Os	osmium	814	2.2	3306	5285	22.587	136
77	Ir	iridium	865	2.2	2719	4701	22.562	132
78	Pt	platinum	864	2.2	2041	4098	21.5	130.
79	Au	gold	890.	2.4	1337	3129	19.3	130.
80	Hg	mercury	1007	1.9	234	630.	13.5336	132
81	Tl	thallium	589	1.8	577	1746	11.8	144
82	Pb	lead	716	1.8	600.	2022	11.3	145
83	Bi	bismuth	703	1.9	544	1837	9.79	150.
84	Po	polonium	812	2.0	527	1235	9.20	142
85	At	astatine	—	2.2	575	—	—	148
86	Rn	radon	1037	—	202	211	0.009074	146
87	Fr	francium	393	0.7	300.	—	—	242
88	Ra	radium	509	0.9	969	—	5	211
89	Ac	actinium	499	1.1	1323	3471	10.	201
Elements 90 and above have been omitted.								

*boiling point at standard pressure
**density of solids and liquids at room temperature and density of gases at 298 K and 101.3 kPa
— no data available
Source: *CRC Handbook for Chemistry and Physics*, 91st ed., 2010–2011, CRC Press

Table T
Important Formulas and Equations

Density	$d = \dfrac{m}{V}$	d = density m = mass V = volume
Mole Calculations	number of moles = $\dfrac{\text{given mass}}{\text{gram-formula mass}}$	
Percent Error	% error = $\dfrac{\text{measured value} - \text{accepted value}}{\text{accepted value}} \times 100$	
Percent Composition	% composition by mass = $\dfrac{\text{mass of part}}{\text{mass of whole}} \times 100$	
Concentration	parts per million = $\dfrac{\text{mass of solute}}{\text{mass of solution}} \times 1\,000\,000$	
	molarity = $\dfrac{\text{moles of solute}}{\text{liter of solution}}$	
Combined Gas Law	$\dfrac{P_1V_1}{T_1} = \dfrac{P_2V_2}{T_2}$	P = pressure V = volume T = temperature
Titration	$M_AV_A = M_BV_B$	M_A = molarity of H^+ M_B = molarity of OH^- V_A = volume of acid V_B = volume of base
Heat	$q = mC\Delta T$ $q = mH_f$ $q = mH_v$	q = heat H_f = heat of fusion m = mass H_v = heat of vaporization C = specific heat capacity ΔT = change in temperature
Temperature	$K = {}^\circ C + 273$	K = kelvin ${}^\circ C$ = degree Celsius

GLOSSARY

absolute zero: The coldest possible temperature; 0K or - 273°C.

acid: 1. An electrolyte that has hydrogen as its positive ion and yields hydrogen ions as the only positive ions in water. 2. A hydrogen ion donor.

activated complex: The temporary, unstable, intermediate union of reactants.

activation energy: The minimum amount of energy needed to produce an activated complex.

activity series: A list of elements in decreasing order of reactivity. An element higher on the list will replace a lower element in a compound in a single replacement (redox) reaction.

addition reaction: A reaction in organic chemistry in which atoms are added to a compound at the site of double or triple bonds.

alcohols: A family of organic compounds. The molecules in this family consist of a hydrocarbon radical combined with one or more hydroxyl (-OH) groups. Their IUPAC name ends in -ol.

aldehydes: A family of organic compounds. The molecules in this family consist of a hydrocarbon radical combined with a terminal - CHO group. Their IUPAC name ends in -al.

alkali metal: A member of Group 1 of the Periodic Table. An active metal having an oxidation value of +1 in compounds.

alkaline metal: Also called an Alkaline Earth metal. A member of Group 2 of the Periodic Table having an oxidation value of + 2 in compounds.

alkanes: The family of saturated hydrocarbons, the molecules of which have the general formula C_nH_{2n+2} and contain only single covalent carbon-to-carbon bonds.

alkenes: A family of unsaturated hydrocarbons. The molecules of this family have the general formula C_nH_{2n} and contain one double covalent carbon-to-carbon bond.

alkyl group: A hydrocarbon radical having the general formula C_nH_{2n+1}.

alkynes: A family of unsaturated hydrocarbons the molecules of which have the general formula C_nH_{2n-2}. These compounds contain one triple covalent carbon-to-carbon bond.

alloy: A homogeneous mixture containing two or more metals.

alpha decay: A transmutation in which a helium nucleus is emitted and the resulting daughter nucleus has an atomic number reduced by 2 and an atomic mass reduced by 4.

alpha particle: A helium nucleus.

amides: Organic compounds having a hydrocarbon chain with an –CONHR on the end of the chain.

amine: A type of organic compound the molecules of which consist of a hydrocarbon radical with a functional –NH_2 attached.

amino acids: Organic compounds with an –NH_2 on an interior carbon and a COOH on the end carbon of the chain. These are the monomers for protein polymers.

analysis: The decomposition of a compound into simpler substances.

anhydrous: Term applied to a hydrate after the water of hydration has been removed.

anion: A negatively charged ion.

anode: The site of oxidation in an electrochemical cell. It is considered to be negatively charged in an electrochemical cell and positively charged in an electrolytic cell.

aqueous: Term applied to a solution in which water is the solvent. It is indicated by (aq), e.g., NaCl (aq).

aromatics: A family of organic compounds. The molecules of this family contain one or more benzene rings.

Arrhenius acid: A substance that yields the hydronium ion in water solution.

Arrhenius base: A substance that contains the hydroxide ion in water solution.

atomic mass: The average mass of the naturally occurring isotopes of an element.

atomic mass units: A relative mass scale with a basic unit $\frac{1}{12}$ the mass of carbon -12.

atomic number: The number of protons in the nucleus of an atom; sometimes designated by the letter Z.

atomic radius: A measure of one-half the distance between two nuclei of an element in the solid phase.

atom: The smallest particle of an element that has the properties of that element.

Avogadro's hypothesis: Equal volumes of gases, when measured under the same conditions of temperature and pressure, contain the same number of molecules.

Avogadro's number: The number of particles in a mole of a compound $- 6.02 \times 10^{23}$.

balanced chemical equation: A chemical equation that conserves both atoms and charge.

base: 1. An electrolyte that yields hydroxide ions in water. Hydroxide is the negative ion in the formula. 2. A hydrogen ion acceptor.

beta particles (rays): High-velocity streams of electrons. Term usually applied to electrons emitted by radioactive atoms or to artificially accelerated electrons (as in the betatron).

binary compound: A compound made up of only two elements.

binding energy: The energy equivalent to the mass defect of an atomic nucleus.

boiling: Turbulence in a liquid caused by the rapid formation of bubbles of vapor at the boiling point of the liquid. See **boiling point**.

boiling point: The temperature at which the vapor pressure of a liquid is equal to the pressure exerted on the liquid. When the atmospheric pressure is 760 torr, the boiling point of water is, by definition, 100.°C, or 373K.

bond strength: A measure of the amount of energy that must be supplied to break a chemical bond.

calorie: A unit of heat. The heat required to raise the temperature of one gram of water one Celsius degree.

calorimeter: An apparatus for measuring the quantity of heat liberated or absorbed during a reaction.

carbohydrate: An organic molecule of sugar and starch with the general formula of $C_n(H_2O)_n$.

catalysis: The change in the rate of a reaction by the presence of a substance which is unchanged at the end of the reaction.

catalytic agent: A substance which alters the rate of a chemical change and which remains unchanged at the end of the reaction, e.g., manganese dioxide in the preparation of oxygen.

cathode: (1) The negative electrode of an electrolytic cell. (2) The electrode at which reduction takes place.

cathode rays: Streams of electrons that emanate from the cathode electrode of a discharge tube.

cation: A positive ion. A cation is attracted to the cathode in an electrolysis, in which the cathode is the negative electrode.

Celsius scale: The temperature scale on which the temperature of freezing water is given the value 0°C, and the temperature of boiling water is 100°C.

chain reaction: A series of reactions in which each reaction is initiated by the energy produced in the preceding reaction.

chemical bond: The linkage between atoms due to the attraction of opposite charges, to the magnetic attractions of shared electrons, or to a combination of such forces.

chemical change: A change in the composition of substances with accompanying changes in properties.

chemical equilibrium: A condition in which two chemical changes exactly oppose each other. Equilibrium is a dynamic condition in which concentrations do not change, and the rate of the forward reaction is equal to the rate of the reverse reaction.

chemical property: A characteristic of a substance that is observed only when the substance undergoes chemical change.

chemistry: The study of the composition, structure, and properties of matter, the changes which matter undergoes, and the energy accompanying these changes.

coefficients: The numbers preceding the formulas in chemical equations, indicating the smallest number of molecules of the substance that may take part in the reaction.

colloidal state: That state of matter in which the particle size lies between 10^{-7} cm and 10^{-4} cm in diameter. The colloidal particle is larger in size than ordinary molecules, but smaller in size than particles which can be seen by the ordinary microscope.

collision theory: Assumes for a reaction to occur, the reactants must collide with each other.

combustion: Any chemical action producing noticeable light and heat.

common ion effect: A displacement of equilibrium brought about by increasing the concentration of one of the ions involved.

complex ions: Charged particles which contain more than one atom. Most complex-ion theory is applied to those complex ions which break down, in chemical action, to smaller units.

compound: A substance of definite composition which may be decomposed into two or more elements by chemical change.

concentrated solution: A solution containing a relatively large amount of solute.

concentration: The quantity of a substance in a given volume, usually expressed as molarity, or percentage.

condensation: The process whereby a gas or vapor changes to a liquid.

conservation of mass, energy, atoms, charge: The total mass, energy, number of atoms, and charge of all the reactants must equal the total of all the products.

covalent bond: A bond indicating a pair of shared electrons.

crystal: A solid with a definite shape, made up of plane faces. The shape is due to the atoms or molecules arranged in a definite repeated pattern.

decay: The spontaneous transmutation of a radioactive nucleus.

decomposition: Applied to a chemical change in which a substance breaks down to form two or more simpler substances.

deposition: The change of state from a gas directly to a solid without becoming a liquid.

density: Mass per unit volume, e.g., grams per cubic centimeter.

destructive distillation: A process in which a substance (usually organic) is heated in the absence of air until decomposition takes place. The products of decomposition (volatile matter) are condensed and collected.

diatomic: Two atoms in a molecule. Usually referred to in the case where both atoms are of the same element.

dihydroxy alcohol: An alcohol whose molecules contain two hydroxy groups per molecule. Antifreeze (ethylene glycol) is such a substance.

dilute: Term applied to a solution that contains a relatively small amount of solute. A process by which the concentration of a solution is reduced.

dipole: A molecule that has an uneven charge distribution. Asymmetrical molecules that contain polar covalent bonds are dipoles.

dissociation: The separation of the ions of an ionic compound, especially during the process of dissolving.

distillate: The product produced by condensation of the vapors produced during distillation.

distillation: The process by which a substance is boiled and the vapors are condensed and recovered.

double bond: The sharing of two pairs of bonding electrons between two atoms.

double replacement: A type of chemical reaction where two reactants are both compounds and there are two compounds for products. The compounds are electrolytes. One of the products is either water or a precipitate.

ductility: That property of a substance which permits its being drawn into wire.

electrical conductivity: Ability to allow electrical charge, as ions in solution or electrons in metals, to move from one place to another.

electrochemical cell: A cell in which the redox reaction is conducted in such a way that the electrons travel through a wire between the substance being oxidized and the substance being reduced.

electrolysis: A chemical reaction that takes place when an electric current is applied to a substance.

electrolyte: A substance whose water solution conducts an electric current.

electron: A fundamental particle of matter having a negative electric charge.

electron-dot symbols: Symbols that contain the symbol of the element and indicate the number of valence electrons.

electronegativity: A measure of the attraction of a nucleus for the electrons in a covalent bond.

electron configuration: A notation that indicates how many electrons are in each energy level around an atom.

electroplating: The process of layering a metal onto a surface in an electrolytic cell.

element: A substance that cannot be decomposed by ordinary chemical means.

empirical formula: A formula showing the simplest ratio of the elements in a chemical compound.

endothermic reaction: A reaction in which the products contain more potential energy than the reactants. ΔH for an endothermic reaction is positive.

end point: That point of a titration when an indicator shows that equivalent amounts of reactants have reacted.

energy: Often defined as the ability to do work. The total amount of energy in a reaction must remain the same. However, it may be changed from one type of energy to another.

enthalpy: A measure of the potential energy of a substance.

entropy: A measure of the amount of randomness of the particles of a substance.

equilibrium: A dynamic chemical condition in which opposing reactions are proceeding at equal rates, producing an apparent constant condition.

equivalency point: (stoichiometric point). The point in an acid-base titration where the moles of hydrogen ions in the acid are equal to the moles of hydroxide ions in the base.

esterification: The reaction of an acid and an alcohol to produce water and an ester.

esters: Organic molecules formed along with water when organic acids react with alcohols. Some fruits and herbs have characteristic odors and flavors due to esters. The general formula of an ester is $R-COO-R^1$, where R and R^1 are hydrocarbons.

ethers: Organic molecules formed along with water from the reaction of two alcohol molecules. The formula is $R-O-R^1$, where R and R^1 are hydrocarbons

evaporation: The changing of a substance from the liquid to the gaseous phase by the absorption of heat.

excited atom: The state of an atom when an electron moves to a higher energy level leaving a lower energy level unfilled. This makes the atom less stable.

exothermic reaction: A reaction in which the products contain less energy than the reactants.

families: The vertical groupings of chemically similar elements in the Periodic Table of the Elements.

fermentation: The production of ethanol and carbon dioxide by the action of enzymes on an organic compound.

filtration: A method used to separate solids from liquids.

fission: A nuclear reaction in which large nuclei are split into smaller nuclear fragments.

formula mass: The sum of all of the atomic masses in a formula; primarily used to describe the mass of ionic substances.

fractional distillation: The separation of different liquids in a mixture by using the different boiling points of the components.

free energy: A measure of the tendency of a reaction to proceed spontaneously. It is represented by ΔG in the Gibbs equation:

$$\Delta G = \Delta H - (T \Delta S).$$

freezing point: The temperature at which both the solid and liquid phases of a substance can exist in equilibrium.

freezing point depression: The lowering of the normal freezing point of a liquid by the addition of solute. One mole of particles lowers the freezing point of 1kg of water by $1.86°C$.

functional group: An atom or group of atoms responsible for specific properties and characteristics of organic compounds.

fusion: 1. The change of a substance from the solid to the liquid phase. 2. A nuclear reaction in which light nuclei combine to form a heavier nucleus.

gamma rays: High energy X-rays emitted from the nucleus of a radioactive element.

gas density: The mass of a liter of gas expressed in grams per liter.

gas phase: The phase of matter which has neither definite volume nor shape.

Gay-Lussac's Law: The volumes of combining gases are in small whole number ratios.

gram-atomic mass: The gram amount of an element numerically equal to the atomic mass of the element.

gram-formula mass: The mass of a substance in grams numerically equal to the formula mass of a compound.

ground state: The condition of an atom in which the electrons occupy the lowest available energy levels.

groups: The vertical columns of the periodic table, also called families.

half-life: The length of time needed for one-half of a given radioactive substance to undergo decay.

half reaction: Either the oxidation or reduction portion of a redox reaction.

halogen: A member of Group 17 of the Periodic Table.

halides: Compounds that have one or more halogen atoms.

halogenation: The placing of a halogen on a carbon chain.

hardness: Ability to resist scratching.

heat: The flow of energy between objects of unequal temperature.

heat of condensation: The amount of heat released as a unit of mass of a substance changes from a vapor to a liquid.

heat of formation: Amount of heat gained or lost during the formation of one mole of a compound from its elements.

heat of fusion: The amount of heat needed to change a unit mass of a substance from solid to liquid phase.

heat of reaction: The amount of heat released or absorbed in a reaction.

heat of solidification: The amount of heat released as a unit mass of a substance changes from the liquid to the solid phase.

heat of vaporization: The amount of heat needed to evaporate a unit mass of a liquid at its boiling point.

heterogeneous: Consisting of different ingredients.

homogeneous: Having similar properties throughout.

homologous series: A group of compounds using the same elements that differ by a constant number of such elements as they become larger.

hydrocarbon: An organic compound which contains only carbon and hydrogen.

hydrogen bond: A bond formed between molecules each having a covalently bonded hydrogen atom to an atom with a high electronegativity value.

hydrogenation: The addition of hydrogen to a substance.

ideal gas: A theoretical gas which occupies no volume and whose particles have no attraction for each other.

insoluble: Does not dissolve readily.

intermolecular forces: the forces of attraction between molecules in the solid or liquid states. They can be weak (see **van der Waal's forces**) or strong (see **hydrogen bonding**) resulting in different physical properties such as melting point and solubility.

ion: A charged atom or group of charged atoms.

ionic bond: A bond formed by the exchange of an electron between two atoms.

ionization energy: The amount of energy needed to remove an electron from a neutral gaseous atom.

isomers: Compounds that have the same molecular formula, but different structural formulas.

isotopes: Nuclei that have the same number of protons but different numbers of neutrons and hence different atomic masses.

IUPAC: International Union of Pure and Applied Chemistry.

IUPAC nomenclature: A system of naming organic compounds as approved by the IUPAC.

Joule: A unit of energy. It takes 4.18 Joules of heat to warm up one gram of water one degree Celsius.

Kelvin scale: Also called the absolute temperature scale. The zero point is the coldest possible temperature. One Kelvin degree is equivalent to one Celsius degree. A Celsius reading can be converted to Kelvin by adding 273 to the Celsius reading. Kelvin temperature designations do not use the degree sign.

kernel: The nucleus and electrons of an atom, except the valence electrons.

ketones: A family of organic compounds whose molecules contain −CHO as the functional group. Propanone (acetone) is the most common member.

kinetic energy: Energy of motion.

kinetic molecular theory: A theory that explains the behavior of gases in terms of the motion of their molecules.

liquid: A phase of matter having a definite volume but taking the shape of its container.

litmus: An indicator that is red in acidic and blue in basic solutions.

malleable: Can be flattened into sheets by hammering. Not brittle.

mass defect: The amount of matter that was converted into energy as protons and neutrons combined to form nuclei.

mass number: The sum of the protons and neutrons in a nucleus.

melting point: The temperature at which a solid melts and can coexist with the liquid phase of that substance. It is the same as freezing point, and involves the addition of potential energy to the substance without a temperature change.

meniscus: The curved surface of a liquid in a container. Instruments are calibrated so that volume readings are taken at the bottom of the meniscus.

matter: Anything that takes up space and has mass.

metals: Atoms that lose electrons in chemical reactions to become positive ions.

metallic bond: The force that holds metallic atoms together in the solid or liquid phase. It is due to attraction between valence electrons and the positive kernels.

metalloid: Members of the Periodic Table that have both metallic and nonmetallic characteristics.

miscible: Term applied to liquid substances that can be mixed to form a solution.

mixture: A substance not having definite proportions and containing two or more components that are not chemically combined.

molar volume: the volume occupied by 1 mole of a gas. This volume is 22.4L at STP; it contains 1 mole of molecules.

molarity: The concentration of a solution expressed as the number of moles of solute per liter of solution.

mole: 6.02×10^{23} particles. This number of particles can be obtained by taking the molecular mass of a substance in grams.

molecular formula: A formula that indicates the number of atoms that are present in the smallest particle that has the chemical properties of the substance.

molecular mass: The sum of the atomic masses of the atoms in a molecular formula.

molecule: The smallest unit of a substance that has the chemical properties of the substance; a discrete particle formed by covalently bonded atoms.

monohydroxy alcohol: An alcohol whose molecules contain only one hydroxyl group.

negative charge: Having attraction to protons. The same charge as electrons. All particles containing this charge will repel each other and attract all positively charged particles.

neutralization: A reaction between an Arrhenius acid and an Arrhenius base to produce a salt and water.

neutron: A neutral nuclear particle having a mass of 1 atomic mass unit (u)

noble gas: Also called an **inert gas**. A member of Group 18 of the Periodic Table. These gases have filled valence electron shells and generally do not enter chemical reactions.

nonelectrolyte: An aqueous solution that does not conduct an electric current.

nonmetals: Elements whose valence electron shells are almost complete.

nonpolar bonds: Covalent bonds in which the two atoms have an equal share of the bonding electrons as measured by their electronegativity values.

nonpolar molecules: Molecules that have a symmetrical shape and a symmetrical charge distribution.

normal boiling point: The boiling point of a liquid at standard pressure.

nuclear fuel: The element used as the source of energy in a nuclear reactor. Often U-235.

nucleons: Particles found in the nucleus. Protons and neutrons are the most commonly identified.

nucleus: The small, dense center of an atom that contains almost all of the mass of the atom in the form of protons and neutrons.

nuclide: any isotope that is identified by symbol, atomic number, and mass number.

octet: A stable configuration of 8 valence electrons.

orbital: The area or space of an atom where an electron of a particular energy content is most likely to be found.

organic acids: Organic compounds the molecules of which contain the carboxyl (–COOH) group as their functional group.

organic chemistry: The chemistry of carbon compounds, particularly hydrocarbons and their derivatives,

ore: A mineral that can be used to produce a metal.

oxidation: The process by which a particle loses an electron, or appears to lose an electron, as indicated by a gain in oxidation number.

oxidation numbers: Values assigned to particles for the purpose of identifying oxidation and reduction processes.

particle accelerators: Devices that accelerate charged particles by magnetic or electric fields.

parts per million (ppm): A unit of concentration. The number of milligrams of solute per liter of solution.

period: A horizontal sequence of elements on the Periodic Table that begins with an alkali metal and ends with a noble (inert) gas. The first period of the Table begins with hydrogen and ends with helium.

Periodic Law: The properties of the elements are a periodic function of their atomic numbers.

peroxide: A binary compound that contains more oxygen atoms than are normally expected. Oxygen is assigned the oxidation value of –1 in such compounds.

pH: A method of expressing the acidity or basicity of a substance on a scale from I to 14. It is the negative of the logarithm of the hydrogen ion concentration.

phase: refers to the gas, liquid, or solid condition of matter.

phenolphthalein: An indicator that is colorless in acidic solutions and pink in basic solutions.

physical change: Those changes in properties that do not result in new chemical compounds.

pOH: The negative of the logarithm of the hydroxide ion concentration.

polar bond: A bond in which the electron pair is shared unequally by the two atoms, resulting in a dipole. The element with the higher electronegativity value is assigned the negative portion of the dipole.

polar molecule: A molecule containing polar bonds with an asymmetric shape and therefore an asymmetric charge distribution.

polyatomic ion: Two or more atoms that are chemically combined and possess a net electric charge, also called a radical.

polymer: A compound with a high molecular mass that consists of many smaller subunits (monomers) that have been bonded together.

polymerization: The process of forming molecules of high molecular mass by the joining of smaller molecules into a chain.

positive charge: Having attraction for electrons. The same charge as protons. All particles containing this charge will repel each other and attract all negatively charged particles.

positron: A fundamental particle with a mass identical to that of an electron, but having a positive charge.

potential energy: Often called stored energy. A particle has potential energy because of its position, phase, or composition.

precipitate: A solid that is formed when two liquids are mixed.

proton: A particle found in the nucleus having a charge of $+1$ and having a mass of 1 atomic mass unit (u).

radiation: Energy emitted from an object.

radioactive dating: A method of determining the age of an object by the use of the half-lives of radioactive elements in the sample.

radioactivity: The spontaneous release of energy by a nucleus.

radioisotope: A radioactive isotope of an element.

reactant: One of the substances consumed in a chemical reaction, a starting substance.

redox: Term used to describe the process in which oxidation and reduction take place.

reduction: The process by which a particle gains electrons, as identified by a decrease in oxidation number.

reversible reaction: A reaction in which the products can reform into the reactants.

salt: An ionic substance consisting of a positive metallic ion and a negative ion other than the hydroxide ion.

salt bridge: A passageway for the movement of ions in an electrochemical cell.

saponification: The reaction of a base plus an ester to produce an alcohol and a soap.

saturated hydrocarbon: A hydrocarbon molecule containing only single covalent carbon-to-carbon bonds.

saturated solution: A solution in which as much solute has been dissolved as is possible for the given temperature.

secondary carbon: A carbon in an organic compound that is directly attached to two other carbon atoms.

single bonds: Covalent bonds between atoms in which one pair of electrons is shared between the atoms.

single replacement: A type of chemical reaction where an element reacts with a compound to produce a different element and compound. This type of reaction is invariably a redox reaction.

solid: The phase of matter whose particles have a definite crystalline arrangement. Solids have both definite volume and a definite shape.

solubility: the maximum amount of a substance that can dissolve in a certain amount of a particular solvent at a particular temperature.

solubility curve: A curve showing the solubility of a solute as a function of temperature.

soluble: Readily dissolves.

solute: The dissolved portion of a solution; the substance present in lesser amount.

solution: A homogeneous mixture of solute in solvent.

solvent: The part of a solution in which the solute is dissolved; the substance present in the greater amount.

specific heat capacity: the amount of energy needed to raise the temperature of one gram of a substance one degree Celsius.

spectrum: The series of lines of radiant energy produced as electrons return from higher to lower energy levels.

spontaneous reaction: A reaction that once begun will continue until one of the reactants has been consumed.

standard conditions: 760 torr and 0°C; referred to as STP.

standard solution: A solution of known concentration.

STP: Standard temperature and pressure.

Stock system: A system of nomenclature in which a Roman numeral is used to show the charge of any metallic ion that can carry different charges.

stoichiometry: The study of the quantitative aspects of formulas and equations.

strong acids: Acids that are highly ionized in solution.

strong bases: Bases that are highly ionized in solution.

structural formula: A molecular formula that shows how the atoms are arranged in the molecule.

sublimation: A change between the solid and gaseous phases without a noticeable liquid phase.

substance: any variety of matter, all specimens of which have identical properties and composition.

substitution: A type of organic reaction where a functional group, often a halogen, replaces a hydrogen on usually a saturated compound.

supersaturated solution: A solution that contains more dissolved solute than would be present in a saturated solution at the same temperature.

symbol: An upper case letter or an upper case letter plus a lower case letter used to represent an atom of an element or one mole of atoms of that element.

synthesis: A type of chemical reaction where a more complex compound is made from two or more simpler substances.

temperature: A measure of the average kinetic energy of the particles of a substance.

ternary acid: An acid containing three different elements per molecule.

thermal energy: the energy associated with the random motion of atoms and molecules.

tincture: A solution in which the solvent is ethanol.

titration: A process in which a solution of known concentration is used to determine the concentration of another solution.

torr: A unit of pressure. Each torr is equivalent to 1 mm Hg. Standard pressure is 760 torr.

tracer: A radioisotope used to follow the path of a chemical reaction.

transition element: An element with an incomplete energy level other than its valence level. Groups 3-11 on the Periodic Table.

transmutation: The conversion of atoms of one element into a different element.

transuranic element: An element with an atomic number greater than 92.

trihydroxy alcohol: An alcohol molecule with three hydroxy groups, such as 1,2,3, propanetriol (glycerol or glycerine).

triple bond: The sharing of three pairs of electrons between two atoms.

tritium: The isotope of hydrogen whose nuclei contain one proton and two neutrons and have a mass of 3 atomic mass units (u)

u: Abbreviation of atomic mass unit.

unsaturated hydrocarbon: A hydrocarbon molecule that contains at least one double or triple bond.

unsaturated solution: A solution in which more solute can be dissolved at a given temperature.

valence electrons: The electrons at the highest principal quantum level, the outermost electrons of an atom.

van der Waals forces: Weak attractive forces between molecules in the solid and liquid phases. They are the result of temporary dipoles in molecules caused by the random, asymmetric motion of electrons.

vapor: The gaseous phase of a substance that is normally a solid or liquid at room temperature.

vapor pressure: The pressure exerted by the vapors of a liquid or a solid.

voltaic cell: An electrochemical cell where the redox reaction occurs spontaneously producing electical energy. Batteries are voltaic cells.

water of hydration: Also known as water of crystallization; the number of water molecules chemically attached to a particle of the substance in the solid state.

wave-mechanical energy: A model of the atom also known as the electron cloud model. Using the wave-like properties of electrons and quantum mechanics, this model puts the electrons in probability distributions called orbitals around the nucleus of the atom.

weak acid: An acid that is only slightly ionized.

weak base: A base that is only slightly ionized.

weak electrolyte: A substance whose water solution is a poor conductor of electricity.

Index

C

D

U

V

W

Z

ADDITIONAL CONSTRUCTED RESPONSE QUESTIONS

1. Wood is made of cellulose, a carbohydrate like starches and sugars with the general formula of approximately $C_x(H_2O)_x$. (Examples of sugars are glucose, $C_6H_{12}O_6$, and sucrose, $C_{12}H_{22}O_{11}$). Freshly cut "green" wood contains 35% water by mass while "dry" wood that has been stacked and dried for a year after being cut is 25% water by mass. It is said that dry wood is a better fuel because it produces more heat than green wood.

   a. Explain why a dry log would burn producing more heat than the same log would have produced when burned green. (1 point)

   _____

   _____

   _____

   _____

   _____

   b. Calculate the heat produced by burning a 10.0 kg green log. (Assume the heat of combustion of all carbohydrates is the same). (3 points)

2.

2 – propanol, $CH_3CHOHCH_3$, commonly called isopropyl alcohol, is rubbing alcohol.

1, 2 – propanediol, $CH_3CHOHCH_2OH$, commonly called propylene glycol, is an antifreeze.

1, 2, 3 – propanetriol, $CH_2OHCHOHCH_2OH$, commonly called glycerol, is a lubricant and sweetener.

All three of these liquids are organic compounds, but, unlike most organic compounds, all readily dissolve in water.

a. Explain why these three compounds dissolve in water. (1 point)

_____

_____

_____

_____

_____

_____

b. Which of these three compounds would have the highest boiling point? Explain your answer. (2 points)

_____

_____

_____

_____

_____

_____

_____

3. Gypsum is hydrated plaster-of-paris, $CaSO_4 \cdot x\ H_2O$.

   a. Make a <u>list</u> of the common high school laboratory equipment that a student must assemble to find the percent water and x in the formula for gypsum. (4 points)

   _____

   _____

   _____

   _____

   _____

   b. Explain how that equipment would be used to answer question (**a.**) above. (2 points)

   _____

   _____

   _____

   _____

   _____

   _____

   _____

   _____

4. A sketch of the Rutherford gold foil experiment is drawn below.

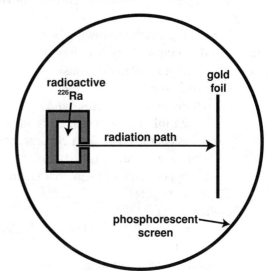

a. Write the nuclear equation for the decay of Ra – 226. (1 point)

_____

b. What is the name of the decay product of Ra – 226 that is detected on the screen? (1 point)

_____

c. Draw and label the path of the radiation stream that is most intense <u>after</u> it strikes the gold foil. (1 point)

d. Draw and label the path of radiation that had a close encounter with a gold nucleus <u>after</u> it strikes the gold foil. (1 point)

e. Sketch the path the radiation from the Ra – 226 would take if the stream were to pass between charged metal plates in the drawing below. (The top plate has negative charge and the bottom is positive.) (1 point)

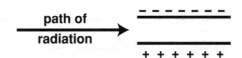

5. In a high school chemistry laboratory, a student wants to find the molar concentration (to 3 significant figures) of 200. mL of HCl (aq) that is known to be 2 ± 0.5 **M**.

The student has the following items:
   – 10.0 grams of dry Na(OH)(s)
   – a 100. mL volumetric flask
   – two 25.0 mL burets
   – a buret holder on a ringstand
   – a shelf full of assorted beakers and Erlenmeyer flasks
   – a wash bottle containing 200. mL of deionized water
   – a balance (accurate to 0.01 g)
   – a dropper bottle of phenolphthalein solution
   – safety goggles
   – lab apron
                    (all glassware is clean)

List the steps the student must follow in their proper order to gather the data. (4 points)

_____

_____

_____

_____

_____

_____

_____

_____

_____

_____

_____

_____

_____

6. Three students, A, B, and C, determined the density of iron, each by a different method. Each repeated the experiment four times and recorded the following results:

Student A	Student B	Student C
7.81 g/cm³	8.312 g/cm³	7.6 g/cm³
8.29	7.562	7.5
7.52	7.601	7.7
7.58	8.532	7.5

a. Which student had the greatest <u>precision</u>? (1 point) _____

b. Which student had the greatest <u>accuracy</u>? (1 point) _____

c. Which student had the greatest number of <u>significant</u> _____
figures? (1 point)

d. Explain how you determined your three answers above. (3 points)

_____

_____

_____

_____

_____

_____

_____

_____

_____

_____

_____

_____

Questions 7 through 9 are based on excerpts from magazine articles. Base your answers to these questions on the passages and your knowledge of chemistry.

Littlest catalysts get a lot of support

Industry has long used reactive metal clusters stuck to larger, inert particles as chemical catalysts. Today, developers of catalysts are making those clusters as small as possible to maximize exposed metal and thus speed up reactions. . . .

In the study, the team observed four-atom clusters of iridium metal catalyzing reactions such as the conversion of mixed propylene and hydrogen into propane gas. The iridium clusters were attached to either aluminum oxide or magnesium oxide particles.

The scientists determined that . . . the propane-producing reaction was 10 times faster on aluminum oxide than on magnesium oxide, another indication that these carriers aren't so inert.

excerpts taken from "Littlest catalysts get a lot of support" by Peter Weiss, Science News, March 2, 2002 Vol. 161, No. 9, p. 141.

---

7. a. Explain why making the catalyst particles as small as possible would speed up the reaction rates. (1 point)

_____

_____

b. Write a balanced molecular equation for the conversion of mixed propylene (propene) and hydrogen gas with iridium metal catalyst into propane gas. (2 points)

c. What kind of organic reaction is answer (b) above? (1 point) _____

d. The potential energy diagram on the right is for the reaction of propene and hydrogen to produce propane with **no** catalyst.

Draw and label the changes to the graph if iridium catalyst on $Al_2O_3(s)$ is used in the reaction. (1 point) Draw and label the changes to the graph if iridium catalyst on MgO(s) is used in the reaction. (1 point)

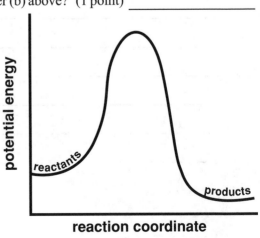

WE'RE MISSING AN ATOM

The periodic table remains one of the few markers of stability in the ever-changing world of scientific theory. So in 1999, it was with much fanfare that Kenneth Gregorich, a nuclear chemist at the Lawrence Berkeley National Laboratory, and his colleagues announced that they had discovered two new elements not in the table. By crashing krypton particles into a chunk of lead, they had created element 118 and the element into which it rapidly decayed, element 116. . . .

It was with less of a flourish that they retracted their claim last July. After two years of futile attempts to reproduce the elusive atoms, Gregorich and his team reanalyzed their data and discovered that there was no "there" there. The decay chains they thought they had seen, from element 118 to 116 to 114 and so on, were nowhere to be found. As far as the periodic table goes, element 118 wasn't around long enough to get a real name: scientists refer to it only by its assigned nomenclature: ununoctium.

excerpts taken from "WE'RE MISSING AN ATOM" by Lauren Gravitz, Discover, January, 2002. p. 43.

---

8. a. Name a <u>physical</u> and a <u>chemical</u> property you would expect of element atomic number 118 (Uuo). (2 points)

physical _____   chemical _____

b. Using krypton –84 and lead –206, write a balanced nuclear equation producing the new element (Uuo) having atomic number 118. (Assume no additional products.) (2 points)

c. When element Uuo (#118), produced in the reaction above, decays into element #116 (Uuh), what single particle is emitted? _____

d. The reaction in the answer to question (b) is known as _____ (1 point)
(1) fusion   (2) fission   (3) artificial transmutation   (4) radioactivity

e. The reaction in the answer to question (b) was attempted in a _____ (1 point)
(1) particle accelerator   (2) Geiger counter   (3) cloud chamber   (4) nuclear reactor

f. What was the flaw in the attempted discovery of element #118? (This flaw is a concept that must be demonstrated by any discovery before it is accepted.) (1 point)

_____

_____

PRACTICE

REGENTS

EXAMINATIONS

Part A

Answer all questions in this part.

Directions (1–30): For *each* statement or question, record on your separate answer sheet the *number* of the word or expression that, of those given, best completes the statement or answers the question. Some questions may require the use of the *2011 Edition Reference Tables for Physical Setting/Chemistry*.

1 Compared to an electron, which particle has a charge that is equal in magnitude but opposite in sign?

(1) an alpha particle   (3) a neutron
(2) a beta particle   (4) a proton

2 The mass of a proton is approximately equal to

(1) 1 atomic mass unit
(2) 12 atomic mass units
(3) the mass of 1 mole of carbon atoms
(4) the mass of 12 moles of electrons

3 Which property *decreases* when the elements in Group 17 are considered in order of increasing atomic number?

(1) atomic mass   (3) melting point
(2) atomic radius   (4) electronegativity

4 Any substance composed of two or more elements that are chemically combined in a fixed proportion is

(1) an isomer   (3) a solution
(2) an isotope   (4) a compound

5 Which term refers to how strongly an atom of an element attracts electrons in a chemical bond with an atom of a different element?

(1) entropy
(2) electronegativity
(3) activation energy
(4) first ionization energy

6 At STP, which substance has metallic bonding?

(1) ammonium chloride (3) iodine
(2) barium oxide   (4) silver

7 What is the number of electrons shared between the carbon atoms in a molecule of ethyne?

(1) 6   (3) 8
(2) 2   (4) 4

8 Which atom in the ground state has a stable valence electron configuration?

(1) Ar   (3) Si
(2) Al   (4) Na

9 What occurs when two fluorine atoms react to produce a fluorine molecule?

(1) Energy is absorbed as a bond is broken.
(2) Energy is absorbed as a bond is formed.
(3) Energy is released as a bond is broken.
(4) Energy is released as a bond is formed.

10 Which gas sample at STP has the same number of molecules as a 2.0-liter sample of $Cl_2(g)$ at STP?

(1) 1.0 L of $NH_3(g)$ (3) 3.0 L of $CO_2(g)$
(2) 2.0 L of $CH_4(g)$ (4) 4.0 L of $NO(g)$

11 All atoms of uranium have the same

(1) mass number
(2) atomic number
(3) number of neutrons plus protons
(4) number of neutrons plus electrons

12 The concentration of a solution can be expressed in

(1) kelvins
(2) milliliters
(3) joules per kilogram
(4) moles per liter

(1)

13 Compared to the boiling point and the freezing point of water at 1 atmosphere, a 1.0 M $CaCl_2(aq)$ solution at 1 atmosphere has a

(1) lower boiling point and a lower freezing point
(2) lower boiling point and a higher freezing point
(3) higher boiling point and a lower freezing point
(4) higher boiling point and higher freezing point

14 According to the kinetic molecular theory, which statement describes an ideal gas?

(1) The gas particles are diatomic.
(2) Energy is created when the gas particles collide.
(3) There are no attractive forces between the gas particles.
(4) The distance between the gas particles is small, compared to their size.

15 Which physical change is endothermic?

(1) $CO_2(s) \rightarrow CO_2(g)$ (3) $CO_2(g) \rightarrow CO_2(\ell)$
(2) $CO_2(\ell) \rightarrow CO_2(s)$ (4) $CO_2(g) \rightarrow CO_2(s)$

16 Which Group 16 element combines with hydrogen to form a compound that has the strongest hydrogen bonding between its molecules?

(1) oxygen   (3) sulfur
(2) selenium   (4) tellurium

17 Hydrocarbons are composed of the elements

(1) carbon and hydrogen, only
(2) carbon and oxygen, only
(3) carbon, hydrogen, and oxygen
(4) carbon, nitrogen, and oxygen

18 Which atom is bonded to the carbon atom in the functional group of a ketone?

(1) fluorine   (3) nitrogen
(2) hydrogen   (4) oxygen

19 Two types of organic reactions are

(1) addition and sublimation
(2) deposition and saponification
(3) decomposition and evaporation
(4) esterification and polymerization

20 The isomers butane and methylpropane have

(1) the same molecular formula and the same properties
(2) the same molecular formula and different properties
(3) different molecular formulas and the same properties
(4) different molecular formulas and different properties

21 In a redox reaction, which particles are lost and gained in equal numbers?

(1) electrons   (3) hydroxide ions
(2) neutrons   (4) hydronium ions

22 What is the oxidation state for a Mn atom?

(1) 0   (3) +3
(2) +7   (4) +4

23 Which compounds are classified as electrolytes?

(1) KNO_3 and H_2SO_4
(2) KNO_3 and CH_3OH
(3) CH_3OCH_3 and H_2SO_4
(4) CH_3OCH_3 and CH_3OH

24 Which compound is an Arrhenius base?

(1) CO_2 (3) $Ca(OH)_2$
(2) $CaSO_4$ (4) C_2H_5OH

25 According to one acid-base theory, a water molecule acts as a base when it accepts

(1) an H^+ ion   (3) a neutron
(2) an OH^- ion   (4) an electron

(2)

26 Given the equation representing a system at equilibrium:

$$N_2(g) + 3H_2(g) \rightleftharpoons 2NH_3(g)$$

Which statement describes this reaction at equilibrium?

(1) The concentration of $N_2(g)$ decreases.
(2) The concentration of $N_2(g)$ is constant.
(3) The rate of the reverse reaction decreases.
(4) The rate of the reverse reaction increases.

27 The acidity or alkalinity of an unknown aqueous solution is indicated by its

(1) pH value
(2) electronegativity value
(3) percent by mass concentration
(4) percent by volume concentration

28 The laboratory process in which the volume of a solution of known concentration is used to determine the concentration of another solution is called

(1) distillation   (3) titration
(2) fermentation   (4) transmutation

29 Which list of nuclear emissions is arranged in order from the greatest penetrating power to the least penetrating power?

(1) alpha particle, beta particle, gamma ray
(2) alpha particle, gamma ray, beta particle
(3) gamma ray, alpha particle, beta particle
(4) gamma ray, beta particle, alpha particle

30 Given the diagram representing a reaction:

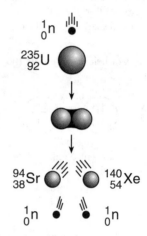

Which type of change is represented?

(1) fission   (3) deposition
(2) fusion   (4) evaporation

Part B–1

Answer all questions in this part.

Directions (31–50): For *each* statement or question, record on your separate answer sheet the *number* of the word or expression that, of those given, best completes the statement or answers the question. Some questions may require the use of the *2011 Edition Reference Tables for Physical Setting/Chemistry*.

31 Which electron shell contains the valence electrons of a radium atom in the ground state?

(1) the sixth shell   (3) the seventh shell
(2) the second shell   (4) the eighteenth shell

32 Each diagram below represents the nucleus of an atom.

How many different elements are represented by the diagrams?

(1) 1   (3) 3
(2) 2   (4) 4

33 Chlorine and element *X* have similar chemical properties. An atom of element *X* could have an electron configuration of

(1) 2-2   (3) 2-8-8
(2) 2-8-1   (4) 2-8-18-7

34 Which group of elements contains a metalloid?

(1) Group 8   (3) Group 16
(2) Group 2   (4) Group 18

35 Which Lewis electron-dot diagram represents a fluoride ion?

   (1)   (2)   (3)   (4)

36 In the formula for the compound XCl_4, the *X* could represent

(1) C   (3) Mg
(2) H   (4) Zn

37 The formula C_2H_4 can be classified as

(1) a structural formula, only
(2) a molecular formula, only
(3) both a structural formula and an empirical formula
(4) both a molecular formula and an empirical formula

38 Given the balanced equation representing a reaction:

$$4Al(s) + 3O_2(g) \rightarrow 2Al_2O_3(s)$$

How many moles of Al(s) react completely with 4.50 moles of $O_2(g)$ to produce 3.00 moles of $Al_2O_3(s)$?

(1) 1.50 mol   (3) 6.00 mol
(2) 2.00 mol   (4) 4.00 mol

39 What is the percent composition by mass of oxygen in $Ca(NO_3)_2$ (gram-formula mass = 164 g/mol)?

(1) 9.8%   (3) 48%
(2) 29%   (4) 59%

40 Given the balanced equation representing a reaction:

$$6Li + N_2 \rightarrow 2Li_3N$$

Which type of chemical reaction is represented by this equation?

(1) synthesis   (3) single replacement
(2) decomposition   (4) double replacement

41 Which elements can react to produce a molecular compound?

(1) calcium and chlorine
(2) hydrogen and sulfur
(3) lithium and fluorine
(4) magnesium and oxygen

42 Compared to a 1.0-mole sample of NaCl(s), a 1.0-mole sample of NaCl(ℓ) has a *different*

(1) number of ions
(2) empirical formula
(3) gram-formula mass
(4) electrical conductivity

43 Which property of an unsaturated solution of sodium chloride in water remains the same when more water is added to the solution?

(1) density of the solution
(2) boiling point of the solution
(3) mass of sodium chloride in the solution
(4) percent by mass of water in the solution

44 Which ion combines with Ba^{2+} to form a compound that is most soluble in water?

(1) S^{2-} (3) CO_3^{2-}
(2) OH^- (4) SO_4^{2-}

45 When a sample of gas is cooled in a sealed, rigid container, the pressure the gas exerts on the walls of the container will decrease because the gas particles hit the walls of the container

(1) less often and with less force
(2) less often and with more force
(3) more often and with less force
(4) more often and with more force

46 A rigid cylinder with a movable piston contains 50.0 liters of a gas at 30.0°C with a pressure of 1.00 atmosphere. What is the volume of the gas in the cylinder at STP?

(1) 5.49 L   (3) 55.5 L
(2) 45.0 L   (4) 455 L

47 Given the potential energy diagram for a chemical reaction:

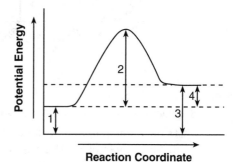

Which numbered interval represents the heat of reaction?

(1) 1   (3) 3
(2) 2   (4) 4

Base your answers to questions 48 and 49 on the graph below and on your knowledge of chemistry.

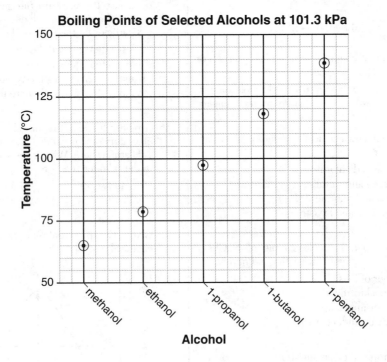

Boiling Points of Selected Alcohols at 101.3 kPa

48 What is represented by the number "1" in the IUPAC name for three of these alcohols?

(1) the number of isomers for each alcohol

(2) the number of $-OH$ groups for each carbon atom in each alcohol molecule

(3) the location of an $-OH$ group on one end of the carbon chain in each alcohol molecule

(4) the location of an $-OH$ group in the middle of the carbon chain in each alcohol molecule

49 What can be concluded from this graph?

(1) At 101.3 kPa, water has a higher boiling point than 1-butanol.

(2) At 101.3 kPa, water has a lower boiling point than ethanol.

(3) The greater the number of carbon atoms per alcohol molecule, the lower the boiling point of the alcohol.

(4) The greater the number of carbon atoms per alcohol molecule, the higher the boiling point of the alcohol.

50 In the laboratory, a student investigates the effect of concentration on the reaction between HCl(aq) and Mg(s), changing only the concentration of HCl(aq). Data for two trials in the investigation are shown in the table below.

Data Table

Trial	Volume of HCl(aq) (mL)	Concentration of HCl(aq) (M)	Mass of Mg(s) (g)	Reaction Time (s)
1	50.0	0.2	0.1	48
2	50.0	0.4	0.1	?

Compared to trial 1, what is the expected reaction time for trial 2 and the explanation for that result?

(1) less than 48 s, because there are fewer effective particle collisions per second
(2) less than 48 s, because there are more effective particle collisions per second
(3) more than 48 s, because there are fewer effective particle collisions per second
(4) more than 48 s, because there are more effective particle collisions per second

Part B–2

Answer all questions in this part.

Directions (51–65): Record your answers in the spaces provided in your answer booklet. Some questions may require the use of the *2011 Edition Reference Tables for Physical Setting/Chemistry*.

51 Determine the volume of 2.00 M HCl(aq) solution required to completely neutralize 20.0 milliliters of 1.00 M NaOH(aq) solution. [1]

52 Determine the mass of KNO_3 that dissolves in 100. grams of water at 40.°C to produce a saturated solution. [1]

53 State, in terms of molecular polarity, why ethanol is soluble in water. [1]

Base your answers to questions 54 through 56 on the information below and on your knowledge of chemistry.

Three elements, represented by *D*, *E*, and *Q*, are located in Period 3. Some properties of these elements are listed in the table below. A student's experimental result indicates that the density of element *Q* is 2.10 g/cm³, at room temperature and standard pressure.

**Properties of Samples of Three Elements
at Room Temperature and Standard Pressure**

Element	Phase	Mass (g)	Density (g/cm³)	Oxide Formula
D	solid	50.0	0.97	D_2O
E	solid	50.0	1.74	EO
Q	solid	50.0	2.00	QO_2 or QO_3

54 Identify the physical property in the table that could be used to differentiate the samples of the three elements from each other. [1]

55 Identify the group on the Periodic Table to which element *D* belongs. [1]

56 Determine the percent error between the student's experimental density and the accepted density of element *Q*. [1]

Base your answers to questions 57 through 59 on the information below and on your knowledge of chemistry.

The equation below represents an equilibrium system of $SO_2(g)$, $O_2(g)$, and $SO_3(g)$. The reaction can be catalyzed by vanadium or platinum.

$$2SO_2(g) + O_2(g) \rightleftharpoons 2SO_3(g) + energy$$

57 Compare the rates of the forward and reverse reactions at equilibrium. [1]

58 State how the equilibrium shifts when $SO_3(g)$ is removed from the system. [1]

59 A potential energy diagram for the forward reaction is shown *in your answer booklet*. On this diagram, draw a dashed line to show how the potential energy changes when the reaction occurs by the catalyzed pathway. [1]

_____

Base your answers to questions 60 and 61 on the information below and on your knowledge of chemistry.

The formulas for two compounds are shown below.

Compound A   **Compound B**

60 Explain, in terms of bonding, why compound A is saturated. [1]

61 Explain, in terms of molecular structure, why the chemical properties of compound A are different from the chemical properties of compound B. [1]

_____

Base your answers to questions 62 through 65 on the information below and on your knowledge of chemistry.

Some isotopes of potassium are K-37, K-39, K-40, K-41, and K-42. The natural abundance and the atomic mass for the naturally occurring isotopes of potassium are shown in the table below.

Naturally Occurring Isotopes of Potassium

Isotope Notation	Natural Abundance (%)	Atomic Mass (u)
K-39	93.26	38.96
K-40	0.01	39.96
K-41	6.73	40.96

62 Identify the decay mode of K-37. [1]

63 Complete the nuclear equation *in your answer booklet* for the decay of K-40 by writing a notation for the missing nuclide. [1]

64 Determine the fraction of an original sample of K-42 that remains unchanged after 24.72 hours. [1]

65 Show a numerical setup for calculating the atomic mass of potassium. [1]

Part C

Answer all questions in this part.

Directions (66–85): Record your answers in the spaces provided in your answer booklet. Some questions may require the use of the *2011 Edition Reference Tables for Physical Setting/Chemistry*.

Base your answers to questions 66 through 68 on the information below and on your knowledge of chemistry.

The Bohr model of the atom was developed in the early part of the twentieth century. A diagram of the Bohr model for one atom, in the ground state, of a specific element is shown below. The nucleus of this atom contains 4 protons and 5 neutrons.

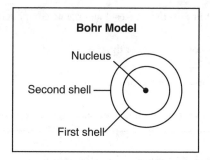

66 State the atomic number and the mass number of this element. [1]

67 State the number of electrons in *each* shell in this atom in the ground state. [1]

68 Using the Bohr model, describe the changes in electron energy and electron location when an atom changes from the ground state to an excited state. [1]

Base your answers to questions 69 through 72 on the information below and on your knowledge of chemistry.

In 1828, Friedrich Wöhler produced urea when he heated a solution of ammonium cyanate. This reaction is represented by the balanced equation below.

Ammonium cyanate   Urea

69 Identify the element in urea that makes it an organic compound. [1]

70 Determine the gram-formula mass of the product. [1]

71 Write an empirical formula for the product. [1]

72 Explain why this balanced equation represents a conservation of atoms. [1]

Base your answers to questions 73 through 75 on the information below and on your knowledge of chemistry.

Rubbing alcohol sold in stores is aqueous 2-propanol, $CH_3CHOHCH_3$(aq). Rubbing alcohol is available in concentrations of 70.% and 91% 2-propanol by volume.

To make 100. mL of 70.% aqueous 2-propanol, 70. mL of 2-propanol is diluted with enough water to produce a total volume of 100. mL. In a laboratory investigation, a student is given a 132-mL sample of 91% aqueous 2-propanol to separate using the process of distillation.

73 State evidence that indicates the proportions of the components in rubbing alcohol can vary. [1]

74 Identify the property of the components that makes it possible to use distillation to separate the 2-propanol from water. [1]

75 Determine the maximum volume of 2-propanol in the 132-mL sample. [1]

Base your answers to questions 76 through 79 on the information below and on your knowledge of chemistry.

A sample of seawater is analyzed. The table below gives the concentration of some ions in the sample.

**Concentration of Some Ions
in a Seawater Sample**

Ion	Concentration (M)
Cl^-	0.545
Na^+	0.468
Mg^{2+}	0.054
SO_4^{2-}	0.028
Ca^{2+}	0.010
K^+	0.010

76 Write a chemical formula of *one* compound formed by the combination of K^+ ions with one of these ions as water completely evaporates from the seawater sample. [1]

77 Determine the number of moles of the SO_4^{2-} ion in a 1400.-liter sample of the seawater. [1]

78 Compare the radius of an Mg^{2+} ion in the seawater to the radius of an Mg atom. [1]

79 Using the key *in your answer booklet*, draw *two* water molecules in the box, showing the orientation of *each* water molecule toward the calcium ion. [1]

_____

Base your answers to questions 80 through 82 on the information below and on your knowledge of chemistry.

A scientist bubbled HCl(g) through a sample of $H_2O(\ell)$. This process is represented by the balanced equation below.

$$H_2O(\ell) + HCl(g) \rightarrow H_3O^+(aq) + Cl^-(aq)$$

The scientist measured the pH of the liquid in the flask before and after the gas was bubbled through the water. The initial pH value of the water was 7.0 and the final pH value of the solution was 3.0.

80 Explain, in terms of ions, why the gaseous reactant in the equation is classified as an Arrhenius acid. [1]

81 What would be the color of bromcresol green if it had been added to the water in the flask before any of the HCl(g) was bubbled through the water? [1]

82 Compare the hydronium ion concentration of the solution that has the pH value of 3.0 to the hydronium ion concentration of the water. [1]

(13)

Base your answers to questions 83 through 85 on the information below and on your knowledge of chemistry.

A small digital clock can be powered by a battery made from two potatoes and some household materials. The "potato clock" battery consists of two cells connected in a way to produce enough electricity to allow the clock to operate. In each cell, zinc atoms react to form zinc ions. Hydrogen ions from phosphoric acid in the potatoes react to form hydrogen gas. The labeled diagram and balanced ionic equation below show the reaction, the materials, and connections necessary to make a "potato clock" battery.

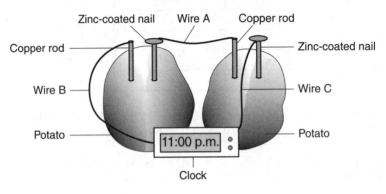

$$Zn(s) + 2H^+(aq) \rightarrow Zn^{2+}(aq) + H_2(g)$$

83 State the direction of electron flow in wire A as the two cells operate. [1]

84 Write a balanced half-reaction equation for the oxidation that occurs in the "potato clock" battery. [1]

85 Explain why phosphoric acid is needed for the battery to operate. [1]

The University of the State of New York

REGENTS HIGH SCHOOL EXAMINATION

PHYSICAL SETTING
CHEMISTRY

PRACTICE EXAM #1

—————

ANSWER BOOKLET

☐ Male

Student . . . . Sex: ☐ Female

Teacher . . . .

School . . . . Grade . . . .

Record your answers for Part B–2 and Part C in this booklet.

Part B–2

51 _____ mL

52 _____ g

53 _____

_____

_____

54 _____

55 _____

56 _____ %

57 _____

_____

_____

58 _____

_____

59

60 _____

_____

_____

61 _____

_____

_____

62 _____

63 $^{40}_{19}K \rightarrow {}^{0}_{-1}e\ +$ _____

64 _____

65

Part C

66 Atomic number: _____

Mass number: _____

67 Number of electrons in first shell: _____

Number of electrons in second shell: _____

68 Change in electron energy: _____

_____

Change in electron location: _____

_____

69 _____

70 _____ **g/mol**

71 _____

72 _____

_____

_____

73 _____

_____

_____

74 _____

_____

_____

75 _____ mL

76 _____

77 _____ mol

78 _____

_____

_____

79

Key

● = hydrogen atom

○ = oxygen atom

● over ○ = water molecule

Ca^{2+}

80 _____

_____

_____

81 _____

82 _____

_____

_____

83 _____

_____

_____

84 _____

85 _____

_____

_____

Part A

Answer all questions in this part.

Directions (1–30): For *each* statement or question, record on your separate answer sheet the *number* of the word or expression that, of those given, best completes the statement or answers the question. Some questions may require the use of the *2011 Edition Reference Tables for Physical Setting/Chemistry*.

1 Which statement describes the charge of an electron and the charge of a proton?

  (1) An electron and a proton both have a charge of +1.

  (2) An electron and a proton both have a charge of −1.

  (3) An electron has a charge of +1, and a proton has a charge of −1.

  (4) An electron has a charge of −1, and a proton has a charge of +1.

2 Which subatomic particles are found in the nucleus of an atom of beryllium?

  (1) electrons and protons

  (2) electrons and positrons

  (3) neutrons and protons

  (4) neutrons and electrons

3 The elements in Period 4 on the Periodic Table are arranged in order of increasing

  (1) atomic radius

  (2) atomic number

  (3) number of valence electrons

  (4) number of occupied shells of electrons

4 Which phrase describes two forms of solid carbon, diamond and graphite, at STP?

  (1) the same crystal structure and the same properties

  (2) the same crystal structure and different properties

  (3) different crystal structures and the same properties

  (4) different crystal structures and different properties

5 Which element has six valence electrons in each of its atoms in the ground state?

  (1) Se   (3) Kr

  (2) As   (4) Ga

6 What is the chemical name for $H_2SO_3(aq)$?

  (1) sulfuric acid

  (2) sulfurous acid

  (3) hydrosulfuric acid

  (4) hydrosulfurous acid

7 Which substance is most soluble in water?

  (1) $(NH_4)_3PO_4$ (3) Ag_2SO_4

  (2) $Cu(OH)_2$ (4) $CaCO_3$

8 Which type of bonding is present in a sample of an element that is malleable?

  (1) ionic   (3) nonpolar covalent

  (2) metallic   (4) polar covalent

9 Which atom has the greatest attraction for the electrons in a chemical bond?

  (1) hydrogen   (3) silicon

  (2) oxygen   (4) sulfur

10 Which type of reaction involves the transfer of electrons?

  (1) alpha decay

  (2) double replacement

  (3) neutralization

  (4) oxidation-reduction

11 A 10.0-gram sample of nitrogen is at STP. Which property will increase when the sample is cooled to 72 K at standard pressure?

  (1) mass   (3) density

  (2) volume   (4) temperature

12 Which element is a gas at STP?

  (1) sulfur   (3) potassium

  (2) xenon   (4) phosphorus

13 A 5.0-gram sample of Fe(s) is to be placed in 100. milliliters of HCl(aq). Which changes will result in the fastest rate of reaction?

(1) increasing the surface area of Fe(s) and increasing the concentration of HCl(aq)

(2) increasing the surface area of Fe(s) and decreasing the concentration of HCl(aq)

(3) decreasing the surface area of Fe(s) and increasing the concentration of HCl(aq)

(4) decreasing the surface area of Fe(s) and decreasing the concentration of HCl(aq)

14 Which process is commonly used to separate a mixture of ethanol and water?

(1) distillation   (3) filtration
(2) ionization   (4) titration

15 A sample of hydrogen gas will behave most like an ideal gas under the conditions of

(1) low pressure and low temperature
(2) low pressure and high temperature
(3) high pressure and low temperature
(4) high pressure and high temperature

16 The collision theory states that a reaction is most likely to occur when the reactant particles collide with the proper

(1) formula masses
(2) molecular masses
(3) density and volume
(4) energy and orientation

17 At STP, which sample contains the same number of molecules as 3.0 liters of $H_2(g)$?

(1) 1.5 L of $NH_3(g)$ (3) 3.0 L of $CH_4(g)$
(2) 2.0 L of $CO_2(g)$ (4) 6.0 L of $N_2(g)$

18 The addition of a catalyst to a chemical reaction provides an alternate pathway that

(1) increases the potential energy of reactants
(2) decreases the potential energy of reactants
(3) increases the activation energy
(4) decreases the activation energy

19 A sample of water is boiling as heat is added at a constant rate. Which statement describes the potential energy and the average kinetic energy of the water molecules in this sample?

(1) The potential energy decreases and the average kinetic energy remains the same.

(2) The potential energy decreases and the average kinetic energy increases.

(3) The potential energy increases and the average kinetic energy remains the same.

(4) The potential energy increases and the average kinetic energy increases.

20 Entropy is a measure of the

(1) acidity of a sample
(2) disorder of a system
(3) concentration of a solution
(4) chemical activity of an element

21 Which element has atoms that can bond with each other to form ring, chain, and network structures?

(1) aluminum   (3) carbon
(2) calcium   (4) argon

22 What is the number of electrons shared in the multiple carbon-carbon bond in one molecule of 1-pentyne?

(1) 6   (3) 3
(2) 2   (4) 8

23 Butanal, butanone, and diethyl ether have different properties because the molecules of each compound differ in their

(1) numbers of carbon atoms
(2) numbers of oxygen atoms
(3) types of functional groups
(4) types of radioactive isotopes

24 What occurs when a magnesium atom becomes a magnesium ion?

(1) Electrons are gained and the oxidation number increases.

(2) Electrons are gained and the oxidation number decreases.

(3) Electrons are lost and the oxidation number increases.

(4) Electrons are lost and the oxidation number decreases.

25 Energy is required to produce a chemical change during

(1) chromatography   (3) boiling

(2) electrolysis   (4) melting

26 The reaction of an Arrhenius acid with an Arrhenius base produces water and

(1) a salt   (3) an aldehyde

(2) an ester   (4) a halocarbon

27 One acid-base theory defines an acid as an

(1) H^- acceptor   (3) H^+ acceptor

(2) H^- donor   (4) H^+ donor

28 Which phrase describes the decay modes and the half-lives of K-37 and K-42?

(1) the same decay mode but different half-lives

(2) the same decay mode and the same half-life

(3) different decay modes and different half-lives

(4) different decay modes but the same half-life

29 Which particle has a mass that is approximately equal to the mass of a proton?

(1) an alpha particle   (3) a neutron

(2) a beta particle   (4) a positron

30 Which change occurs during a nuclear fission reaction?

(1) Covalent bonds are converted to ionic bonds.

(2) Isotopes are converted to isomers.

(3) Temperature is converted to mass.

(4) Matter is converted to energy.

Part B–1

Answer all questions in this part.

Directions (31–50): For *each* statement or question, record on your separate answer sheet the *number* of the word or expression that, of those given, best completes the statement or answers the question. Some questions may require the use of the *2011 Edition Reference Tables for Physical Setting/Chemistry*.

31 Which notations represent hydrogen isotopes?

(1) $_1^1H$ and $_1^2H$ (3) $_2^1H$ and $_3^1H$

(2) $_1^1H$ and $_2^4H$ (4) $_1^2H$ and $_2^7H$

32 Naturally occurring gallium is a mixture of isotopes that contains 60.11% of Ga-69 (atomic mass = 68.93 u) and 39.89% of Ga-71 (atomic mass = 70.92 u). Which numerical setup can be used to determine the atomic mass of naturally occurring gallium?

(1) $\dfrac{(68.93\ u + 70.92\ u)}{2}$

(2) $\dfrac{(68.93\ u)(0.6011)}{(70.92\ u)(0.3989)}$

(3) $(68.93\ u)(0.6011) + (70.92\ u)(0.3989)$

(4) $(68.93\ u)(39.89) + (70.92\ u)(60.11)$

33 Which list of symbols represents nonmetals, only?

(1) B, Al, Ga   (3) C, Si, Ge

(2) Li, Be, B   (4) P, S, Cl

34 In the formula XSO_4, the symbol X could represent the element

(1) Al   (3) Mg

(2) Ar   (4) Na

35 What is the chemical formula for lead(IV) oxide?

(1) PbO_2 (3) Pb_2O

(2) PbO_4 (4) Pb_4O

36 Which statement describes the general trends in electronegativity and atomic radius as the elements in Period 2 are considered in order from left to right?

(1) Both electronegativity and atomic radius increase.

(2) Both electronegativity and atomic radius decrease.

(3) Electronegativity increases and atomic radius decreases.

(4) Electronegativity decreases and atomic radius increases.

37 What is the percent composition by mass of nitrogen in $(NH_4)_2CO_3$ (gram-formula mass = 96.0 g/mol)?

(1) 14.6%   (3) 58.4%

(2) 29.2%   (4) 87.5%

38 Given the balanced equation:

$$2KI + F_2 \rightarrow 2KF + I_2$$

Which type of chemical reaction does this equation represent?

(1) synthesis

(2) decomposition

(3) single replacement

(4) double replacement

39 Which formula represents a nonpolar molecule containing polar covalent bonds?

H–H   O=C=O

(1)   (2)   (3)   (4)

40 A reaction reaches equilibrium at 100.°C. The equation and graph representing this reaction are shown below.

$$N_2O_4(g) \rightleftharpoons 2NO_2(g)$$

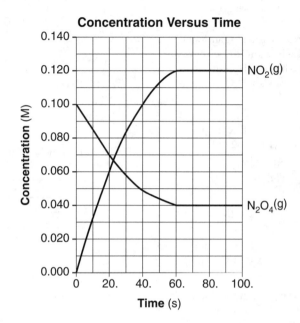

The graph shows that the reaction is at equilibrium after 60. seconds because the concentrations of both $NO_2(g)$ and $N_2O_4(g)$ are

(1) increasing   (3) constant
(2) decreasing   (4) zero

41 Given the balanced equation representing a reaction:

$$2H_2O + energy \rightarrow 2H_2 + O_2$$

Which statement describes the changes in energy and bonding for the reactant?

(1) Energy is absorbed as bonds in H_2O are formed.
(2) Energy is absorbed as bonds in H_2O are broken.
(3) Energy is released as bonds in H_2O are formed.
(4) Energy is released as bonds in H_2O are broken.

42 At standard pressure, what is the temperature at which a saturated solution of NH_4Cl has a concentration of 60. g NH_4Cl/100. g H_2O?

(1) 66°C   (3) 22°C
(2) 57°C   (4) 17°C

43 Which aqueous solution has the highest boiling point at standard pressure?

(1) 1.0 M KCl(aq)   (3) 2.0 M KCl(aq)
(2) 1.0 M $CaCl_2$(aq)   (4) 2.0 M $CaCl_2$(aq)

44 Given the equation representing a system at equilibrium:

$$KNO_3(s) + energy \underset{}{\overset{H_2O}{\rightleftharpoons}} K^+(aq) + NO_3^-(aq)$$

Which change causes the equilibrium to shift?

(1) increasing pressure
(2) increasing temperature
(3) adding a noble gas
(4) adding a catalyst

45 Which hydrocarbon is saturated?

(1) C_2H_2 (3) C_4H_6
(2) C_3H_4 (4) C_4H_{10}

46 Which volume of 0.600 M H_2SO_4(aq) exactly neutralizes 100. milliliters of 0.300 M $Ba(OH)_2$(aq)?

(1) 25.0 mL   (3) 100. mL
(2) 50.0 mL   (4) 200. mL

47 Given the formula for an organic compound:

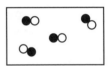

What is the name given to the group in the box?

(1) butyl   (3) methyl
(2) ethyl   (4) propyl

48 Given the particle diagram:

Key
O = atom of an element
● = atom of a different element

Which type of matter is represented by the particle diagram?

(1) an element
(2) a compound
(3) a homogeneous mixture
(4) a heterogeneous mixture

49 Which substance is an electrolyte?

(1) O_2 (3) C_3H_8
(2) Xe   (4) KNO_3

50 Which type of organic reaction produces both water and carbon dioxide?

(1) addition   (3) esterification
(2) combustion   (4) fermentation

Part B–2

Answer all questions in this part.

Directions (51–65): Record your answers in the spaces provided in your answer booklet. Some questions may require the use of the *2011 Edition Reference Tables for Physical Setting/Chemistry*.

51 Draw a Lewis electron-dot diagram for a chloride ion, Cl^-. [1]

Base your answers to questions 52 and 53 on the information below and on your knowledge of chemistry.

At STP, Cl_2 is a gas and I_2 is a solid. When hydrogen reacts with chlorine, the compound hydrogen chloride is formed. When hydrogen reacts with iodine, the compound hydrogen iodide is formed.

52 Balance the equation *in your answer booklet* for the reaction between hydrogen and chlorine, using the smallest whole-number coefficients. [1]

53 Explain, in terms of intermolecular forces, why iodine is a solid at STP but chlorine is a gas at STP. [1]

Base your answers to questions 54 and 55 on the information below and on your knowledge of chemistry.

Some properties of the element sodium are listed below.
• is a soft, silver-colored metal
• melts at a temperature of 371 K
• oxidizes easily in the presence of air
• forms compounds with nonmetallic elements in nature
• forms sodium chloride in the presence of chlorine gas

54 Identify *one* chemical property of sodium from this list. [1]

55 Convert the melting point of sodium to degrees Celsius. [1]

Base your answers to questions 56 through 58 on the information below and on your knowledge of chemistry.

At standard pressure, water has unusual properties that are due to both its molecular structure and intermolecular forces. For example, although most liquids contract when they freeze, water expands, making ice less dense than liquid water. Water has a much higher boiling point than most other molecular compounds having a similar gram-formula mass.

56 Explain why $H_2O(s)$ floats on $H_2O(\ell)$ when both are at 0°C. [1]

57 State the type of intermolecular force responsible for the unusual boiling point of $H_2O(\ell)$ at standard pressure. [1]

58 Determine the total amount of heat, in joules, required to completely vaporize a 50.0-gram sample of $H_2O(\ell)$ at its boiling point at standard pressure. [1]

Base your answers to questions 59 and 60 on the information below and on your knowledge of chemistry.

At 1023 K and 1 atm, a 3.00-gram sample of $SnO_2(s)$ (gram-formula mass = 151 g/mol) reacts with hydrogen gas to produce tin and water, as shown in the balanced equation below.

$$SnO_2(s) + 2H_2(g) \rightarrow Sn(\ell) + 2H_2O(g)$$

59 Show a numerical setup for calculating the number of moles of $SnO_2(s)$ in the 3.00-gram sample. [1]

60 Determine the number of moles of $Sn(\ell)$ produced when 4.0 moles of $H_2(g)$ is completely consumed. [1]

Base your answers to questions 61 and 62 on the information below and on your knowledge of chemistry.

The incomplete data table below shows the pH value of solutions A and B and the hydrogen ion concentration of solution A.

Hydrogen Ion and pH Data for HCl(aq) Solutions

HCl(aq) Solution	Hydrogen Ion Concentration (M)	pH
A	1.0×10^{-2}	2.0
B	?	5.0

61 State the color of methyl orange in a sample of solution A. [1]

62 Determine the hydrogen ion concentration of solution B. [1]

Base your answers to questions 63 through 65 on the information below and on your knowledge of chemistry.

A sample of helium gas is placed in a rigid cylinder that has a movable piston. The volume of the gas is varied by moving the piston, while the temperature is held constant at 273 K. The volumes and corresponding pressures for three trials are measured and recorded in the data table below. For each of these trials, the product of pressure and volume is also calculated and recorded. For a fourth trial, only the volume is recorded.

**Pressure and Volume Data for
a Sample of Helium Gas at 273 K**

Trial Number	Pressure (atm)	Volume (L)	P x V (L•atm)
1	1.000	0.412	0.412
2	0.750	0.549	0.412
3	0.600	0.687	0.412
4	?	1.373	?

63 State evidence found in the data table that allows the product of pressure and volume for the fourth trial to be predicted. [1]

64 Determine the pressure of the helium gas in trial 4. [1]

65 Compare the average distances between the helium atoms in trial 1 to the average distances between the helium atoms in trial 3. [1]

Part C

Answer all questions in this part.

Directions (66–85): Record your answers in the spaces provided in your answer booklet. Some questions may require the use of the *2011 Edition Reference Tables for Physical Setting/Chemistry*.

Base your answers to questions 66 through 69 on the information below and on your knowledge of chemistry.

Potassium phosphate, K_3PO_4, is a source of dietary potassium found in a popular cereal. According to the Nutrition-Facts label shown on the boxes of this brand of cereal, the accepted value for a one-cup serving of this cereal is 170. milligrams of potassium. The minimum daily requirement of potassium is 3500 milligrams for an adult human.

66 Identify *two* types of chemical bonding in the source of dietary potassium in this cereal. [1]

67 Identify the noble gas whose atoms have the same electron configuration as a potassium ion. [1]

68 Compare the radius of a potassium ion to the radius of a potassium atom. [1]

69 The mass of potassium in a one-cup serving of this cereal is determined to be 172 mg. Show a numerical setup for calculating the percent error for the mass of potassium in this serving. [1]

_____

Base your answers to questions 70 and 71 on the information below and on your knowledge of chemistry.

During photosynthesis, plants use carbon dioxide, water, and light energy to produce glucose, $C_6H_{12}O_6$, and oxygen. The reaction for photosynthesis is represented by the balanced equation below.

$$6CO_2 + 6H_2O + \text{light energy} \rightarrow C_6H_{12}O_6 + 6O_2$$

70 Write the empirical formula for glucose. [1]

71 State evidence that indicates photosynthesis is an endothermic reaction. [1]

_____

Base your answers to questions 72 through 74 on the information below and on your knowledge of chemistry.

Fireworks that contain metallic salts such as sodium, strontium, and barium can generate bright colors. A technician investigates what colors are produced by the metallic salts by performing flame tests. During a flame test, a metallic salt is heated in the flame of a gas burner. Each metallic salt emits a characteristic colored light in the flame.

72 Explain why the electron configuration of 2-7-1-1 represents a sodium atom in an excited state. [1]

73 Explain, in terms of electrons, how a strontium salt emits colored light. [1]

74 State how bright-line spectra viewed through a spectroscope can be used to identify the metal ions in the salts used in the flame tests. [1]

Base your answers to questions 75 through 77 on the information below and on your knowledge of chemistry.

The unique odors and flavors of many fruits are primarily due to small quantities of a certain class of organic compounds. The equation below represents the production of one of these compounds.

| Reactant 1 | Reactant 2 | Product 1 | Product 2 |

75 Show a numerical setup for calculating the gram-formula mass for reactant 1. [1]

76 Explain, in terms of molecular polarity, why reactant 2 is soluble in water. [1]

77 State the class of organic compounds to which product 1 belongs. [1]

Base your answers to questions 78 through 81 on the information below and on your knowledge of chemistry.

A student develops the list shown below that includes laboratory equipment and materials for constructing a voltaic cell.

Laboratory Equipment and Materials
- a strip of zinc
- a strip of copper
- a 250-mL beaker containing 150 mL of 0.1 M zinc nitrate
- a 250-mL beaker containing 150 mL of 0.1 M copper(II) nitrate
- wires
- a voltmeter
- a switch
- a salt bridge

78 State the purpose of the salt bridge in the voltaic cell. [1]

79 Complete and balance the half-reaction equation *in your answer booklet* for the oxidation of the Zn(s) that occurs in the voltaic cell. [1]

80 Compare the activities of the two metals used by the student for constructing the voltaic cell. [1]

81 Identify *one* item of laboratory equipment required to build an electrolytic cell that is *not* included in the list. [1]

Base your answers to questions 82 through 85 on the information below and on your knowledge of chemistry.

In 1896, Antoine H. Becquerel discovered that a uranium compound could expose a photographic plate wrapped in heavy paper in the absence of light. It was shown that the uranium compound was spontaneously releasing particles and high-energy radiation. Further tests showed the emissions from the uranium that exposed the photographic plate were *not* deflected by charged plates.

82 Identify the highly penetrating radioactive emission that exposed the photographic plates. [1]

83 Complete the nuclear equation *in your answer booklet* for the alpha decay of U-238. [1]

84 Determine the number of neutrons in an atom of U-233. [1]

85 Identify the type of nuclear reaction that occurs when an alpha or a beta particle is spontaneously emitted by a radioactive isotope. [1]

The University of the State of New York

REGENTS HIGH SCHOOL EXAMINATION

PHYSICAL SETTING
CHEMISTRY

PRACTICE EXAM #2

ANSWER BOOKLET

Student . . . .   Sex: ☐ Male
                                                                                                                                ☐ Female

Teacher . . . .

School . . . .   Grade . . . .

Record your answers for Part B–2 and Part C in this booklet.

Part B–2

51

52 _____ $H_2(g)$ + _____ $Cl_2(g)$ → _____ $HCl(g)$

53 _____

_____

_____

54 _____

_____

55 _____ °C

56 _____

_____

_____

57 _____

58 _____ J

59

60 _____ mol

61 _____

62 _____ M

63 _____

_____

_____

64 _____ atm

65 _____

_____

_____

Part C

66 _____ and _____

67 _____

68 _____

_____

69

70 _____

71 _____

_____

_____

72 _____

_____

_____

73 _____

_____

_____

74 _____

_____

_____

75

76 _____

_____

_____

77 _____

78 _____

_____

79 $Zn(s) \rightarrow$ _____ + _____

80 _____

_____

_____

81 _____

82 _____

83 $^{238}_{92}U \rightarrow {}^{4}_{2}He +$ _____

84 _____

85 _____

NOTES

Part A

Answer all questions in this part.

Directions (1–30): For *each* statement or question, record on your separate answer sheet the *number* of the word or expression that, of those given, best completes the statement or answers the question. Some questions may require the use of the *2011 Edition Reference Tables for Physical Setting/Chemistry.*

1 Which statement describes the structure of an atom?

(1) The nucleus contains positively charged electrons.
(2) The nucleus contains negatively charged protons.
(3) The nucleus has a positive charge and is surrounded by negatively charged electrons.
(4) The nucleus has a negative charge and is surrounded by positively charged electrons.

2 Which term is defined as the region in an atom where an electron is most likely to be located?

(1) nucleus   (3) quanta
(2) orbital   (4) spectra

3 What is the number of electrons in an atom of scandium?

(1) 21   (3) 45
(2) 24   (4) 66

4 Which particle has the *least* mass?

(1) a proton   (3) a helium atom
(2) an electron   (4) a hydrogen atom

5 Which electron transition in an excited atom results in a release of energy?

(1) first shell to the third shell
(2) second shell to the fourth shell
(3) third shell to the fourth shell
(4) fourth shell to the second shell

6 On the Periodic Table, the number of protons in an atom of an element is indicated by its

(1) atomic mass
(2) atomic number
(3) selected oxidation states
(4) number of valence electrons

7 Which type of formula shows an element symbol for each atom and a line for each bond between atoms?

(1) ionic   (3) empirical
(2) structural   (4) molecular

8 What is conserved during all chemical reactions?

(1) charge   (3) vapor pressure
(2) density   (4) melting point

9 In which type of reaction can two compounds exchange ions to form two different compounds?

(1) synthesis
(2) decomposition
(3) single replacement
(4) double replacement

10 At STP, two 5.0-gram solid samples of different ionic compounds have the same density. These solid samples could be differentiated by their

(1) mass   (3) temperature
(2) volume   (4) solubility in water

11 What is the number of electrons shared between the atoms in an I_2 molecule?

(1) 7   (3) 8
(2) 2   (4) 4

12 Which substance has nonpolar covalent bonds?

(1) Cl_2 (3) SiO_2
(2) SO_3 (4) CCl_4

13 Compared to a potassium atom, a potassium ion has

(1) a smaller radius   (3) fewer protons
(2) a larger radius   (4) more protons

14 Which form of energy is associated with the random motion of particles in a gas?

(1) chemical   (3) nuclear
(2) electrical   (4) thermal

15 The average kinetic energy of water molecules *decreases* when

(1) $H_2O(\ell)$ at 337 K changes to $H_2O(\ell)$ at 300. K
(2) $H_2O(\ell)$ at 373 K changes to $H_2O(g)$ at 373 K
(3) $H_2O(s)$ at 200. K changes to $H_2O(s)$ at 237 K
(4) $H_2O(s)$ at 273 K changes to $H_2O(\ell)$ at 273 K

16 The joule is a unit of

(1) concentration (3) pressure
(2) energy   (4) volume

17 Compared to a sample of helium at STP, the same sample of helium at a higher temperature and a lower pressure

(1) condenses to a liquid
(2) is more soluble in water
(3) forms diatomic molecules
(4) behaves more like an ideal gas

18 A sample of a gas is in a sealed, rigid container that maintains a constant volume. Which changes occur between the gas particles when the sample is heated?

(1) The frequency of collisions increases, and the force of collisions decreases.
(2) The frequency of collisions increases, and the force of collisions increases.
(3) The frequency of collisions decreases, and the force of collisions decreases.
(4) The frequency of collisions decreases, and the force of collisions increases.

19 At STP, which gaseous sample has the same number of molecules as 3.0 liters of $N_2(g)$?

(1) 6.0 L of $F_2(g)$ (3) 3.0 L of $H_2(g)$
(2) 4.5 L of $N_2(g)$ (4) 1.5 L of $Cl_2(g)$

20 Distillation of crude oil from various parts of the world yields different percentages of hydrocarbons. Which statement explains these different percentages?

(1) Each component in a mixture has a different solubility in water.
(2) Hydrocarbons are organic compounds.
(3) The carbons in hydrocarbons may be bonded in chains or rings.
(4) The proportions of components in a mixture can vary.

21 In which 1.0-gram sample are the particles arranged in a crystal structure?

(1) $CaCl_2(s)$ (3) $CH_3OH(\ell)$
(2) $C_2H_6(g)$ (4) $CaI_2(aq)$

22 When a reversible reaction is at equilibrium, the concentration of products and the concentration of reactants must be

(1) decreasing   (3) constant
(2) increasing   (4) equal

23 In chemical reactions, the difference between the potential energy of the products and the potential energy of the reactants is equal to the

(1) activation energy
(2) ionization energy
(3) heat of reaction
(4) heat of vaporization

24 What occurs when a catalyst is added to a chemical reaction?

(1) an alternate reaction pathway with a lower activation energy
(2) an alternate reaction pathway with a higher activation energy
(3) the same reaction pathway with a lower activation energy
(4) the same reaction pathway with a higher activation energy

25 What is the name of the compound with the formula $CH_3CH_2CH_2NH_2$?

(1) 1-propanol   (3) propanal
(2) 1-propanamine   (4) propanamide

26 Which compound is an isomer of $C_2H_5OC_2H_5$?

   (1) CH_3COOH (3) $C_3H_7COCH_3$

   (2) $C_2H_5COOCH_3$ (4) C_4H_9OH

27 Ethanoic acid and 1-butanol can react to produce water and a compound classified as an

   (1) aldehyde   (3) ester

   (2) amide   (4) ether

28 During an oxidation-reduction reaction, the number of electrons gained is

   (1) equal to the number of electrons lost

   (2) equal to the number of protons gained

   (3) less than the number of electrons lost

   (4) less than the number of protons gained

29 Which process requires energy for a nonspontaneous redox reaction to occur?

   (1) deposition   (3) alpha decay

   (2) electrolysis   (4) chromatography

30 Which pair of compounds represents one Arrhenius acid and one Arrhenius base?

   (1) CH_3OH and NaOH   (3) HNO_3 and NaOH

   (2) CH_3OH and HCl   (4) HNO_3 and HCl

Part B–1

Answer all questions in this part.

Directions (31–50): For *each* statement or question, record on your separate answer sheet the *number* of the word or expression that, of those given, best completes the statement or answers the question. Some questions may require the use of the *2011 Edition Reference Tables for Physical Setting/Chemistry.*

31 Which electron configuration represents the electrons of an atom of neon in an excited state?

(1) 2-7   (3) 2-7-1
(2) 2-8   (4) 2-8-1

32 Some information about the two naturally occurring isotopes of gallium is given in the table below.

Natural Abundance of Two Gallium Isotopes

Isotope	Natural Abundance (%)	Atomic Mass (u)
Ga-69	60.11	68.926
Ga-71	39.89	70.925

Which numerical setup can be used to calculate the atomic mass of gallium?

(1) (0.6011)(68.926 u) + (0.3989)(70.925 u)
(2) (60.11)(68.926 u) + (39.89)(70.925 u)
(3) (0.6011)(70.925 u) + (0.3989)(68.926 u)
(4) (60.11)(70.925 u) + (39.89)(68.926 u)

33 A student measures the mass and volume of a sample of copper at room temperature and 101.3 kPa. The mass is 48.9 grams and the volume is 5.00 cubic centimeters. The student calculates the density of the sample. What is the percent error of the student's calculated density?

(1) 7.4%   (3) 9.2%
(2) 8.4%   (4) 10.2%

34 What is the chemical formula for sodium sulfate?

(1) Na_2SO_4 (3) $NaSO_4$
(2) Na_2SO_3 (4) $NaSO_3$

35 Given the balanced equation representing a reaction:

$$2Na(s) + Cl_2(g) \rightarrow 2NaCl(s) + energy$$

If 46 grams of Na and 71 grams of Cl_2 react completely, what is the total mass of NaCl produced?

(1) 58.5 g   (3) 163 g
(2) 117 g   (4) 234 g

36 Given the balanced equation representing a reaction:

$$2NO + O_2 \rightarrow 2NO_2 + energy$$

The mole ratio of NO to NO_2 is

(1) 1 to 1   (3) 3 to 2
(2) 2 to 1   (4) 5 to 2

37 The particle diagram below represents a solid sample of silver.

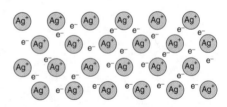

Which type of bonding is present when valence electrons move within the sample?

(1) metallic bonding   (3) covalent bonding
(2) hydrogen bonding   (4) ionic bonding

38 Given the formula representing a molecule:

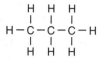

Which statement explains why the molecule is nonpolar?

(1) Electrons are shared between the carbon atoms and the hydrogen atoms.
(2) Electrons are transferred from the carbon atoms to the hydrogen atoms.
(3) The distribution of charge in the molecule is symmetrical.
(4) The distribution of charge in the molecule is asymmetrical.

39 A solid sample of a compound and a liquid sample of the same compound are each tested for electrical conductivity. Which test conclusion indicates that the compound is ionic?

(1) Both the solid and the liquid are good conductors.
(2) Both the solid and the liquid are poor conductors.
(3) The solid is a good conductor, and the liquid is a poor conductor.
(4) The solid is a poor conductor, and the liquid is a good conductor.

40 Which statement explains why 10.0 mL of a 0.50 M H_2SO_4(aq) solution exactly neutralizes 5.0 mL of a 2.0 M NaOH(aq) solution?

(1) The moles of H^+(aq) equal the moles of OH^-(aq).
(2) The moles of H_2SO_4(aq) equal the moles of NaOH(aq).
(3) The moles of H_2SO_4(aq) are greater than the moles of NaOH(aq).
(4) The moles of H^+(aq) are greater than the moles of OH^-(aq).

41 Which particle diagram represents *one* substance in the gas phase?

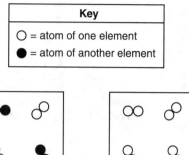

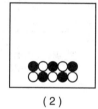

(1)

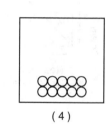

(3)

(2)   (4)

42 Given the equation representing a chemical reaction at equilibrium in a sealed, rigid container:

$$H_2(g) + I_2(g) + energy \rightleftharpoons 2HI(g)$$

When the concentration of $H_2(g)$ is increased by adding more hydrogen gas to the container at constant temperature, the equilibrium shifts

(1) to the right, and the concentration of HI(g) decreases
(2) to the right, and the concentration of HI(g) increases
(3) to the left, and the concentration of HI(g) decreases
(4) to the left, and the concentration of HI(g) increases

43 Which diagram represents the potential energy changes during an exothermic reaction?

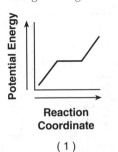

(1)

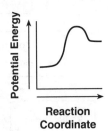

(3)

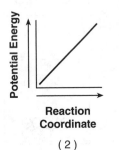

(2)

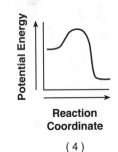

(4)

44 Which compound is classified as an ether?

(1) CH_3CHO (3) CH_3COCH_3
(2) CH_3OCH_3 (4) CH_3COOCH_3

45 Given the equation representing a reversible reaction:

$$HCO_3^-(aq) + H_2O(\ell) \rightleftharpoons H_2CO_3(aq) + OH^-(aq)$$

Which formula represents the H^+ acceptor in the forward reaction?

(1) $HCO_3^-(aq)$ (3) $H_2CO_3(aq)$
(2) $H_2O(\ell)$ (4) $OH^-(aq)$

46 What is the mass of an original 5.60-gram sample of iron-53 that remains unchanged after 25.53 minutes?

(1) 0.35 g   (3) 1.40 g
(2) 0.70 g   (4) 2.80 g

47 Given the equation representing a nuclear reaction:

$$^1_1H + X \rightarrow {}^6_3Li + {}^4_2He$$

The particle represented by X is

(1) ^9_4Li (3) $^{10}_5Be$
(2) ^9_4Be (4) $^{10}_6C$

48 Fission and fusion reactions both release energy. However, only fusion reactions

(1) require elements with large atomic numbers
(2) create radioactive products
(3) use radioactive reactants
(4) combine light nuclei

49 The chart below shows the crystal shapes and melting points of two forms of solid phosphorus.

Two Forms of Phosphorus

Form of Phosphorus	Crystal Shape	Melting Point (°C)
white	cubic 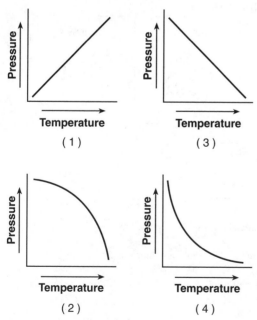	44
black	orthorhombic	610

Which phrase describes the two forms of phosphorus?

(1) same crystal structure and same properties
(2) same crystal structure and different properties
(3) different crystal structures and different properties
(4) different crystal structures and same properties

50 Which graph shows the relationship between pressure and Kelvin temperature for an ideal gas at constant volume?

Pressure

Temperature

(1)

Pressure

Temperature

(3)

Pressure

Temperature

(2)

Pressure

Temperature

(4)

Part B–2

Answer all questions in this part.

Directions (51–65): Record your answers in the spaces provided in your answer booklet. Some questions may require the use of the *2011 Edition Reference Tables for Physical Setting/Chemistry.*

Base your answers to questions 51 through 53 on the information below and on your knowledge of chemistry.

The elements in Group 17 are called halogens. The word "halogen" is derived from Greek and means "salt former."

51 State the trend in electronegativity for the halogens as these elements are considered in order of increasing atomic number.   [1]

52 Identify the type of chemical bond that forms when potassium reacts with bromine to form a salt.   [1]

53 Based on Table *F*, identify *one* ion that reacts with iodide ions in an aqueous solution to form an insoluble compound.   [1]

Base your answers to questions 54 through 57 on the information below and on your knowledge of chemistry.

The diagrams below represent four different atomic nuclei.

Four Atomic Nuclei

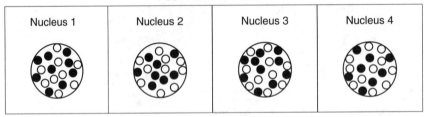

Key
● = proton
○ = neutron

54 Identify the element that has atomic nuclei represented by nucleus 1.   [1]

55 Determine the mass number of the nuclide represented by nucleus 2.   [1]

56 Explain why nucleus 2 and nucleus 4 represent the nuclei of two different isotopes of the same element.   [1]

57 Identify the nucleus above that is found in an atom that has a stable valence electron configuration.   [1]

Base your answers to questions 58 through 60 on the information below and on your knowledge of chemistry.

The equation below represents a chemical reaction at 1 atm and 298 K.

$$2H_2(g) + O_2(g) \rightarrow 2H_2O(g)$$

58 State the change in energy that occurs in order to break the bonds in the hydrogen molecules. [1]

59 In the space *in your answer booklet,* draw a Lewis electron-dot diagram for a water molecule. [1]

60 Compare the strength of attraction for electrons by a hydrogen atom to the strength of attraction for electrons by an oxygen atom within a water molecule. [1]

_____

Base your answers to questions 61 through 63 on the information below and on your knowledge of chemistry.

- A test tube contains a sample of solid stearic acid, an organic acid.
- Both the sample and the test tube have a temperature of 22.0°C.
- The stearic acid melts after the test tube is placed in a beaker with 320. grams of water at 98.0°C.
- The temperature of the liquid stearic acid and water in the beaker reaches 74.0°C.

61 Identify the element in stearic acid that makes it an organic compound. [1]

62 State the direction of heat transfer between the test tube and the water when the test tube was placed in the water. [1]

63 Show a numerical setup for calculating the amount of thermal energy change for the water in the beaker. [1]

_____

Base your answers to questions 64 and 65 on the information below and on your knowledge of chemistry.

A nuclear reaction is represented by the equation below.

$$^3_1H \rightarrow ^3_2He + ^{0}_{-1}e$$

64 Identify the decay mode of hydrogen-3. [1]

65 Explain why the equation represents a transmutation. [1]

_____

Part C

Answer all questions in this part.

Directions (66–85): Record your answers in the spaces provided in your answer booklet. Some questions may require the use of the *2011 Edition Reference Tables for Physical Setting/Chemistry.*

Base your answers to questions 66 through 68 on the information below and on your knowledge of chemistry.

A technician recorded data for two properties of Period 3 elements. The data are shown in the table below.

Two Properties of Period 3 Elements

Element	Na	Mg	Al	Si	P	S	Cl	Ar
Ionic Radius (pm)	95	66	51	41	212	184	181	—
Reaction with Cold Water	reacts vigorously	reacts very slowly	no observable reaction	no observable reaction	no observable reaction	no observable reaction	reacts slowly	no observable reaction

66 Identify the element in this table that is classified as a metalloid. [1]

67 State the phase of chlorine at 281 K and 101.3 kPa. [1]

68 State evidence from the technician's data which indicates that sodium is more active than aluminum. [1]

Base your answers to questions 69 through 71 on the information below and on your knowledge of chemistry.

Ammonia, $NH_3(g)$, can be used as a substitute for fossil fuels in some internal combustion engines. The reaction between ammonia and oxygen in an engine is represented by the unbalanced equation below.

$$NH_3(g) + O_2(g) \rightarrow N_2(g) + H_2O(g) + energy$$

69 Balance the equation *in your answer booklet* for the reaction of ammonia and oxygen, using the smallest whole-number coefficients. [1]

70 Show a numerical setup for calculating the mass, in grams, of a 4.2-mole sample of O_2. Use 32 g/mol as the gram-formula mass of O_2. [1]

71 Determine the new pressure of a 6.40-L sample of oxygen gas at 300. K and 100. kPa after the gas is compressed to 2.40 L at 900. K. [1]

(10)

Base your answers to questions 72 through 76 on the information below and on your knowledge of chemistry.

Fruit growers in Florida protect oranges when the temperature is near freezing by spraying water on them. It is the freezing of the water that protects the oranges from frost damage. When $H_2O(\ell)$ at 0°C changes to $H_2O(s)$ at 0°C, heat energy is released. This energy helps to prevent the temperature inside the orange from dropping below freezing, which could damage the fruit. After harvesting, oranges can be exposed to ethene gas, C_2H_4, to improve their color.

72 Write the empirical formula for ethene. [1]

73 Explain, in terms of bonding, why the hydrocarbon ethene is classified as unsaturated. [1]

74 Determine the gram-formula mass of ethene. [1]

75 Explain, in terms of particle arrangement, why the entropy of the water *decreases* when the water freezes. [1]

76 Determine the quantity of heat released when 2.00 grams of $H_2O(\ell)$ freezes at 0°C. [1]

Base your answers to questions 77 through 80 on the information below and on your knowledge of chemistry.

A student constructs an electrochemical cell during a laboratory investigation. When the switch is closed, electrons flow through the external circuit. The diagram and ionic equation below represent this cell and the reaction that occurs.

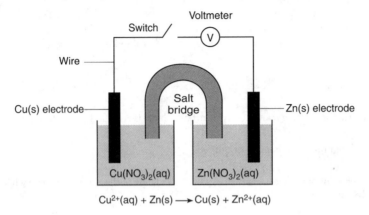

$$Cu^{2+}(aq) + Zn(s) \longrightarrow Cu(s) + Zn^{2+}(aq)$$

77 State the form of energy that is converted to electrical energy in the operating cell. [1]

78 State, in terms of the Cu(s) electrode and the Zn(s) electrode, the direction of electron flow in the external circuit when the cell operates. [1]

79 Write a balanced equation for the half-reaction that occurs in the Cu half-cell when the cell operates. [1]

80 State what happens to the mass of the Cu electrode and the mass of the Zn electrode in the operating cell. [1]

Base your answers to questions 81 and 82 on the information below and on your knowledge of chemistry.

A solution is made by dissolving 70.0 grams of $KNO_3(s)$ in 100. grams of water at 50.°C and standard pressure.

81 Show a numerical setup for calculating the percent by mass of KNO_3 in the solution. [1]

82 Determine the number of additional grams of KNO_3 that must dissolve to make this solution saturated. [1]

Base your answers to questions 83 through 85 on the information below and on your knowledge of chemistry.

Vinegar is a commercial form of acetic acid, $HC_2H_3O_2(aq)$. One sample of vinegar has a pH value of 2.4.

83 Explain, in terms of particles, why $HC_2H_3O_2(aq)$ can conduct an electric current. [1]

84 State the color of bromthymol blue indicator in a sample of the commercial vinegar. [1]

85 State the pH value of a sample that has ten times *fewer* hydronium ions than an equal volume of a vinegar sample with a pH value of 2.4. [1]

The University of the State of New York

REGENTS HIGH SCHOOL EXAMINATION

PHYSICAL SETTING
CHEMISTRY

PRACTICE EXAM #3

_____

ANSWER BOOKLET

☐ Male

Student . . . .   Sex: ☐ Female

Teacher . . . .

School . . . .   Grade . . . .

Record your answers for Part B–2 and Part C in this booklet.

Part B–2
51 _____
_____
_____
52 _____
53 _____

54 _____

55 _____

56 _____

_____

_____

57 _____

58 _____

_____

_____

59

60 _____

_____

_____

61 _____

62 _____

_____

_____

63

64 _____

65 _____

_____

_____

Part C

66 _____

67 _____

68 _____

_____

_____

69 _____ $NH_3(g)$ + _____ $O_2(g) \rightarrow$ _____ $N_2(g)$ + _____ $H_2O(g)$ + energy

70

71 _____ **kPa**

72 _____

73 _____

_____

_____

74 _____ g/mol

75 _____

_____

_____

76 _____ J

77 _____

78 _____

_____

_____

79 _____

80 Cu electrode: _____

Zn electrode: _____

81

82 _____ g

83 _____

_____

_____

84 _____

85 _____

Part A

Answer all questions in this part.

Directions (1–30): For *each* statement or question, record on your separate answer sheet the *number* of the word or expression that, of those given, best completes the statement or answers the question. Some questions may require the use of the *2011 Edition Reference Tables for Physical Setting/Chemistry*.

1 Which statement describes the charge and location of an electron in an atom?

   (1) An electron has a positive charge and is located outside the nucleus.
   (2) An electron has a positive charge and is located in the nucleus.
   (3) An electron has a negative charge and is located outside the nucleus.
   (4) An electron has a negative charge and is located in the nucleus.

2 Which statement explains why a xenon atom is electrically neutral?

   (1) The atom has fewer neutrons than electrons.
   (2) The atom has more protons than electrons.
   (3) The atom has the same number of neutrons and electrons.
   (4) The atom has the same number of protons and electrons.

3 If two atoms are isotopes of the same element, the atoms must have

   (1) the same number of protons and the same number of neutrons
   (2) the same number of protons and a different number of neutrons
   (3) a different number of protons and the same number of neutrons
   (4) a different number of protons and a different number of neutrons

4 Which electrons in a calcium atom in the ground state have the greatest effect on the chemical properties of calcium?

   (1) the two electrons in the first shell
   (2) the two electrons in the fourth shell
   (3) the eight electrons in the second shell
   (4) the eight electrons in the third shell

5 The weighted average of the atomic masses of the naturally occuring isotopes of an element is the

   (1) atomic mass of the element
   (2) atomic number of the element
   (3) mass number of each isotope
   (4) formula mass of each isotope

6 Which element is classified as a metalloid?

   (1) Cr   (3) Sc
   (2) Cs   (4) Si

7 Which statement describes a chemical property of iron?

   (1) Iron oxidizes.
   (2) Iron is a solid at STP.
   (3) Iron melts.
   (4) Iron is attracted to a magnet.

8 Graphite and diamond are two forms of the same element in the solid phase that differ in their

   (1) atomic numbers
   (2) crystal structures
   (3) electronegativities
   (4) empirical formulas

9 Which ion has the largest radius?

   (1) Br^- (3) F^-
   (2) Cl^- (4) I^-

10 Carbon monoxide and carbon dioxide have

   (1) the same chemical properties and the same physical properties
   (2) the same chemical properties and different physical properties
   (3) different chemical properties and the same physical properties
   (4) different chemical properties and different physical properties

11 Based on Table S, which group on the Periodic Table has the element with the highest electronegativity?

(1) Group 1   (3) Group 17
(2) Group 2   (4) Group 18

12 What is represented by the chemical formula $PbCl_2(s)$?

(1) a substance
(2) a solution
(3) a homogeneous mixture
(4) a heterogeneous mixture

13 What is the vapor pressure of propanone at 50.°C?

(1) 37 kPa   (3) 83 kPa
(2) 50. kPa   (4) 101 kPa

14 Which statement describes the charge distribution and the polarity of a CH_4 molecule?

(1) The charge distribution is symmetrical and the molecule is nonpolar.
(2) The charge distribution is asymmetrical and the molecule is nonpolar.
(3) The charge distribution is symmetrical and the molecule is polar.
(4) The charge distribution is asymmetrical and the molecule is polar.

15 In a laboratory investigation, a student separates colored compounds obtained from a mixture of crushed spinach leaves and water by using paper chromatography. The colored compounds separate because of differences in

(1) molecular polarity
(2) malleability
(3) boiling point
(4) electrical conductivity

16 Which phrase describes the motion and attractive forces of ideal gas particles?

(1) random straight-line motion and no attractive forces
(2) random straight-line motion and strong attractive forces
(3) random curved-line motion and no attractive forces
(4) random curved-line motion and strong attractive forces

17 At which temperature will $Hg(\ell)$ and $Hg(s)$ reach equilibrium in a closed system at 1.0 atmosphere?

(1) 234 K   (3) 373 K
(2) 273 K   (4) 630. K

18 A molecule of any organic compound has at least one

(1) ionic bond   (3) oxygen atom
(2) double bond   (4) carbon atom

19 A chemical reaction occurs when reactant particles

(1) are separated by great distances
(2) have no attractive forces between them
(3) collide with proper energy and proper orientation
(4) convert chemical energy into nuclear energy

20 Systems in nature tend to undergo changes toward

(1) lower energy and lower entropy
(2) lower energy and higher entropy
(3) higher energy and lower entropy
(4) higher energy and higher entropy

21 Which formula can represent an alkyne?

(1) C_2H_4 (3) C_3H_4
(2) C_2H_6 (4) C_3H_6

22 Given the formula representing a compound:

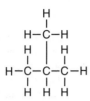

Which formula represents an isomer of this compound?

(1)   (3)

(2)   (4)

23 Which energy conversion occurs in an operating voltaic cell?

(1) chemical energy to electrical energy
(2) chemical energy to nuclear energy
(3) electrical energy to chemical energy
(4) electrical energy to nuclear energy

24 Which process requires energy to decompose a substance?

(1) electrolysis   (3) sublimation
(2) neutralization   (4) synthesis

25 The concentration of which ion is increased when LiOH is dissolved in water?

(1) hydroxide ion   (3) hydronium ion
(2) hydrogen ion   (4) halide ion

26 Which equation represents neutralization?

(1) $6Li(s) + N_2(g) \rightarrow 2Li_3N(s)$

(2) $2Mg(s) + O_2(g) \rightarrow 2MgO(s)$

(3) $2KOH(aq) + H_2SO_4(aq) \rightarrow$
    $K_2SO_4(aq) + 2H_2O(\ell)$

(4) $Pb(NO_3)_2(aq) + K_2CrO_4(aq) \rightarrow$
    $2KNO_3(aq) + PbCrO_4(s)$

27 The stability of an isotope is related to its ratio of

(1) neutrons to positrons
(2) neutrons to protons
(3) electrons to positrons
(4) electrons to protons

28 Which particle has the *least* mass?

(1) alpha particle   (3) neutron
(2) beta particle   (4) proton

29 The energy released during a nuclear reaction is a result of

(1) breaking chemical bonds
(2) forming chemical bonds
(3) mass being converted to energy
(4) energy being converted to mass

30 The use of uranium-238 to determine the age of a geological formation is a beneficial use of

(1) nuclear fusion
(2) nuclear fission
(3) radioactive isomers
(4) radioactive isotopes

Part B–1

Answer all questions in this part.

Directions (31–50): For *each* statement or question, record on your separate answer sheet the *number* of the word or expression that, of those given, best completes the statement or answers the question. Some questions may require the use of the *2011 Edition Reference Tables for Physical Setting/Chemistry*.

Base your answers to questions 31 and 32 on your knowledge of chemistry and the bright-line spectra produced by four elements and the spectrum of a mixture of elements represented in the diagram below.

Bright-Line Spectra

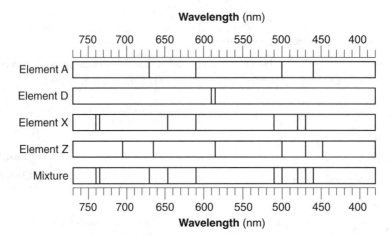

31 Which elements are present in this mixture?

(1) D and A   (3) X and A
(2) D and Z   (4) X and Z

32 Each line in the spectra represents the energy

(1) absorbed as an atom loses an electron
(2) absorbed as an atom gains an electron
(3) released as an electron moves from a lower energy state to a higher energy state
(4) released as an electron moves from a higher energy state to a lower energy state

33 The table below shows the number of protons, neutrons, and electrons in four ions.

Four Ions

Ion	Number of Protons	Number of Neutrons	Number of Electrons
A	8	10	10
E	9	10	10
G	11	12	10
J	12	12	10

Which ion has a charge of 2−?

(1) A   (3) G
(2) E   (4) J

34 What is the approximate mass of an atom that contains 26 protons, 26 electrons and 19 neutrons?

(1) 26 u   (3) 52 u
(2) 45 u   (4) 71 u

35 Which electron configuration represents a potassium atom in an excited state?

(1) 2-7-6   (3) 2-8-8-1
(2) 2-8-5   (4) 2-8-7-2

36 What is the total number of neutrons in an atom of K-42?

(1) 19   (3) 23
(2) 20   (4) 42

37 Given the equation representing a reaction:

$$2C + 3H_2 \rightarrow C_2H_6$$

What is the number of moles of C that must completely react to produce 2.0 moles of C_2H_6?

(1) 1.0 mol   (3) 3.0 mol
(2) 2.0 mol   (4) 4.0 mol

38 Given the equation representing a reaction:

$$Mg(s) + 2HCl(aq) \rightarrow MgCl_2(aq) + H_2(g)$$

Which type of chemical reaction is represented by the equation?

(1) synthesis
(2) decomposition
(3) single replacement
(4) double replacement

(5)

39 The table below lists properties of selected elements at room temperature.

Properties of Selected Elements at Room Temperature

Element	Density (g/cm³)	Malleability	Conductivity
sodium	0.97	yes	good
gold	19.3	yes	good
iodine	4.933	no	poor
tungsten	19.3	yes	good

Based on this table, which statement describes how two of these elements can be differentiated from each other?

(1) Gold can be differentiated from tungsten based on density.
(2) Gold can be differentiated from sodium based on malleability.
(3) Sodium can be differentiated from tungsten based on conductivity.
(4) Sodium can be differentiated from iodine based on malleability.

40 Which particle diagram represents a mixture?

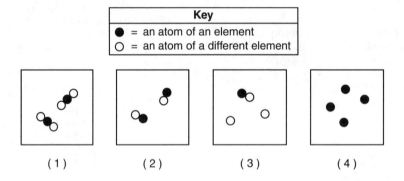

Key
● = an atom of an element
○ = an atom of a different element

(1)   (2)   (3)   (4)

41 An atom of which element reacts with an atom of hydrogen to form a bond with the greatest degree of polarity?

(1) carbon   (3) nitrogen
(2) fluorine   (4) oxygen

42 What is the concentration of an aqueous solution that contains 1.5 moles of NaCl in 500. milliliters of this solution?

(1) 0.30 M   (3) 3.0 M
(2) 0.75 M   (4) 7.5 M

43 The table below shows data for the temperature, pressure, and volume of four gas samples.

Data for Four Gases

Gas Sample	Temperature (K)	Pressure (atm)	Volume (L)
I	600.	2.0	5.0
II	300.	1.0	10.0
III	600.	3.0	5.0
IV	300.	1.0	10.0

Which two gas samples contain the same number of molecules?

(1) I and II   (3) II and III
(2) I and III   (4) II and IV

44 Based on Table I, what is the ΔH value for the production of 1.00 mole of $NO_2(g)$ from its elements at 101.3 kPa and 298 K?

(1) +33.2 kJ   (3) +132.8 kJ
(2) −33.2 kJ   (4) −132.8 kJ

45 Which equation represents an addition reaction?

(1) $C_3H_8 + Cl_2 \rightarrow C_3H_7Cl + HCl$
(2) $C_3H_6 + Cl_2 \rightarrow C_3H_6Cl_2$
(3) $CaCl_2 + Na_2CO_3 \rightarrow CaCO_3 + 2NaCl$
(4) $CaCO_3 \rightarrow CaO + CO_2$

46 Given the balanced equation representing a reaction:

$$Ni(s) + 2HCl(aq) \rightarrow NiCl_2(aq) + H_2(g)$$

In this reaction, each Ni atom

(1) loses 1 electron   (3) gains 1 electron
(2) loses 2 electrons   (4) gains 2 electrons

47 Which equation represents a reduction half-reaction?

(1) $Fe \rightarrow Fe^{3+} + 3e^-$ (3) $Fe^{3+} \rightarrow Fe + 3e^-$
(2) $Fe + 3e^- \rightarrow Fe^{3+}$ (4) $Fe^{3+} + 3e^- \rightarrow Fe$

48 Given the balanced ionic equation representing a reaction:

$$Cu(s) + 2Ag^+(aq) \rightarrow Cu^{2+}(aq) + 2Ag(s)$$

During this reaction, electrons are transferred from

(1) $Cu(s)$ to $Ag^+(aq)$
(2) $Cu^{2+}(aq)$ to $Ag(s)$
(3) $Ag(s)$ to $Cu^{2+}(aq)$
(4) $Ag^+(aq)$ to $Cu(s)$

49 Which metal reacts spontaneously with Sr^{2+} ions?

(1) $Ca(s)$ (3) $Cs(s)$
(2) $Co(s)$ (4) $Cu(s)$

50 Given the balanced equation representing a reaction:

$$HCl + H_2O \rightarrow H_3O^+ + Cl^-$$

The water molecule acts as a base because it

(1) donates an H^+ (3) donates an OH^-
(2) accepts an H^+ (4) accepts an OH^-

Part B–2

Answer all questions in this part.

Directions (51–65): Record your answers in the spaces provided in your answer booklet. Some questions may require the use of the *2011 Edition Reference Tables for Physical Setting/Chemistry.*

51 State the general trend in first ionization energy as the elements in Period 3 are considered from left to right. [1]

52 Identify a type of strong intermolecular force that exists between water molecules, but does *not* exist between carbon dioxide molecules. [1]

53 Draw a structural formula for 2-butanol. [1]

Base your answers to questions 54 through 56 on the information below and on your knowledge of chemistry.

Some compounds of silver are listed with their chemical formulas in the table below.

Silver Compounds

Name	Chemical Formula
silver carbonate	Ag_2CO_3
silver chlorate	$AgClO_3$
silver chloride	$AgCl$
silver sulfate	Ag_2SO_4

54 Explain, in terms of element classification, why silver chloride is an ionic compound. [1]

55 Show a numerical setup for calculating the percent composition by mass of silver in silver carbonate (gram-formula mass = 276 g/mol). [1]

56 Identify the silver compound in the table that is most soluble in water. [1]

Base your answers to questions 57 through 59 on the information below and on your knowledge of chemistry.

When a cobalt-59 atom is bombarded by a subatomic particle, a radioactive cobalt-60 atom is produced. After 21.084 years, 1.20 grams of an original sample of cobalt-60 produced remains unchanged.

57 Complete the nuclear equation by writing a notation for the missing particle. [1]

58 Based on Table *N*, identify the decay mode of cobalt-60. [1]

59 Determine the mass of the original sample of cobalt-60 produced. [1]

Base your answers to questions 60 through 62 on the information below and on your knowledge of chemistry.

A sample of a molecular substance starting as a gas at 206°C and 1 atm is allowed to cool for 16 minutes. This process is represented by the cooling curve below.

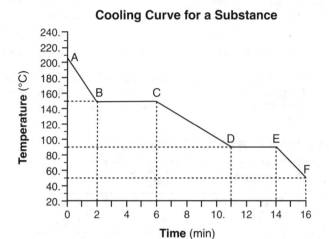

60 Determine the number of minutes that the substance was in the liquid phase, only. [1]

61 Compare the strength of the intermolecular forces within this substance at 180.°C to the strength of the intermolecular forces within this substance at 120.°C. [1]

62 Describe what happens to the potential energy and the average kinetic energy of the molecules in the sample during interval *DE*. [1]

Base your answers to questions 63 through 65 on the information below and on your knowledge of chemistry.

The diagram below represents a cylinder with a moveable piston containing 16.0 g of $O_2(g)$. At 298 K and 0.500 atm, the $O_2(g)$ has a volume of 24.5 liters.

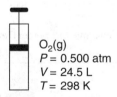

63 Determine the number of moles of $O_2(g)$ in the cylinder. The gram-formula mass of $O_2(g)$ is 32.0 g/mol. [1]

64 State the changes in *both* pressure and temperature of the gas in the cylinder that would increase the frequency of collisions between the $O_2(g)$ molecules. [1]

65 Show a numerical setup for calculating the volume of $O_2(g)$ in the cylinder at 265 K and 1.00 atm. [1]

Part C

Answer all questions in this part.

Directions (66–85): Record your answers in the spaces provided in your answer booklet. Some questions may require the use of the *2011 Edition Reference Tables for Physical Setting/Chemistry*.

Base your answers to questions 66 through 69 on the information below and on your knowledge of chemistry.

In the late 1800s, Dmitri Mendeleev developed a periodic table of the elements known at that time. Based on the pattern in his periodic table, he was able to predict properties of some elements that had not yet been discovered. Information about two of these elements is shown in the table below.

Some Element Properties Predicted by Mendeleev

Predicted Elements	Property	Predicted Value	Actual Value
eka-aluminum (Ea)	density at STP	5.9 g/cm³	5.91 g/cm³
	melting point	low	30.°C
	oxide formula	Ea_2O_3	
	approximate molar mass	68 g/mol	
eka-silicon (Es)	density at STP	5.5 g/cm³	5.3234 g/cm³
	melting point	high	938°C
	oxide formula	EsO_2	
	approximate molar mass	72 g/mol	

66 Identify the phase of Ea at 310. K. [1]

67 Write a chemical formula for the compound formed between Ea and Cl. [1]

68 Identify the element that Mendeleev called eka-silicon, Es. [1]

69 Show a numerical setup for calculating the percent error of Mendeleev's predicted density of Es. [1]

Base your answers to questions 70 through 73 on the information below and your knowledge of chemistry.

Methanol can be manufactured by a reaction that is reversible. In the reaction, carbon monoxide gas and hydrogen gas react using a catalyst. The equation below represents this system at equilibrium.

$$CO(g) + 2H_2(g) \rightleftharpoons CH_3OH(g) + energy$$

70 State the class of organic compounds to which the product of the forward reaction belongs. [1]

71 Compare the rate of the forward reaction to the rate of the reverse reaction in this equilibrium system. [1]

72 Explain, in terms of collision theory, why increasing the concentration of $H_2(g)$ in this system will increase the concentration of $CH_3OH(g)$. [1]

73 State the effect on the rates of both the forward and reverse reactions if no catalyst is used in the system. [1]

Base your answers to questions 74 through 76 on the information below and on your knowledge of chemistry.

Fatty acids, a class of compounds found in living things, are organic acids with long hydrocarbon chains. Linoleic acid, an unsaturated fatty acid, is essential for human skin flexibility and smoothness. The formula below represents a molecule of linoleic acid.

```
     H  H  H  H  H  H  H  H  H  H  H  H  H  H  H  H  H  O
     |  |  |  |  |  |  |  |  |  |  |  |  |  |  |  |  |  ||
 H − C− C− C− C− C− C=C− C− C=C− C− C− C− C− C− C− C− C− O−H
     |  |  |  |  |     |        |  |  |  |  |  |  |
     H  H  H  H  H     H        H  H  H  H  H  H  H
```

74 Write the molecular formula of linoleic acid. [1]

75 Identify the type of chemical bond between the oxygen atom and the hydrogen atom in the linoleic acid molecule. [1]

76 On the diagram *in your answer booklet*, circle the organic acid functional group. [1]

Base your answers to questions 77 through 79 on the information below and on your knowledge of chemistry.

Fuel cells are voltaic cells. In one type of fuel cell, oxygen gas, $O_2(g)$, reacts with hydrogen gas, $H_2(g)$, producing water vapor, $H_2O(g)$, and electrical energy. The unbalanced equation for this redox reaction is shown below.

$$H_2(g) + O_2(g) \rightarrow H_2O(g) + energy$$

A diagram of the fuel cell is shown below. During operation of the fuel cell, hydrogen gas is pumped into one compartment and oxygen gas is pumped into the other compartment. Each compartment has an inner wall that is a porous carbon electrode through which ions flow. Aqueous potassium hydroxide, $KOH(aq)$, and the porous electrodes serve as the salt bridge.

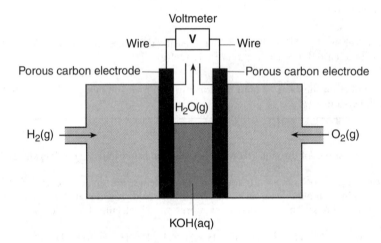

77 Balance the equation *in your answer booklet* for the reaction in this fuel cell, using the smallest whole-number coefficients. [1]

78 Determine the change in oxidation number for oxygen in this operating fuel cell. [1]

79 State the number of moles of electrons that are gained when 5.0 moles of electrons are lost in this reaction. [1]

Base your answers to questions 80 through 82 on the information below and on your knowledge of chemistry.

In a laboratory investigation, a student compares the concentration and pH value of each of four different solutions of hydrochloric acid, HCl(aq), as shown in the table below.

Data for HCl(aq) Solutions

Solution	Concentration of HCl(aq) (M)	pH Value
W	1.0	0
X	0.10	1
Y	0.010	2
Z	0.0010	3

80 State the number of significant figures used to express the concentration of solution Z. [1]

81 Determine the concentration of an HCl(aq) solution that has a pH value of 4. [1]

82 Determine the volume of 0.25 M NaOH(aq) that would exactly neutralize 75.0 milliliters of solution X. [1]

Base your answers to questions 83 through 85 on the information below and on your knowledge of chemistry.

Carbon dioxide is slightly soluble in seawater. As carbon dioxide levels in the atmosphere increase, more CO_2 dissolves in seawater, making the seawater more acidic because carbonic acid, $H_2CO_3(aq)$, is formed.

Seawater also contains aqueous calcium carbonate, $CaCO_3(aq)$, which is used by some marine organisms to make their hard exoskeletons. As the acidity of the sea water changes, the solubility of $CaCO_3$ also changes, as shown in the graph below.

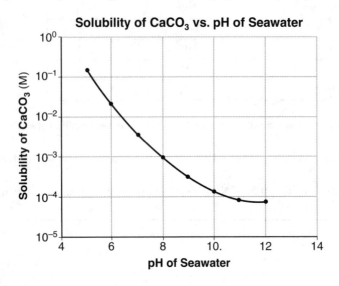

83 State the trend in the solubility of $CaCO_3$ as seawater becomes more acidic. [1]

84 State the color of bromcresol green in a sample of seawater in which the $CaCO_3$ solubility is 10^{-2} M. [1]

85 A sample of seawater has a pH of 8. Determine the new pH of the sample if the hydrogen ion concentration is increased by a factor of 100. [1]

The University of the State of New York

REGENTS HIGH SCHOOL EXAMINATION

PHYSICAL SETTING
CHEMISTRY

PRACTICE EXAM #4

ANSWER BOOKLET

Student . . . .

Teacher . . . .

School . . . . Grade . . . .

Record your answers for Part B–2 and Part C in this booklet.

Part B–2
51
52
53

54 _____

_____

_____

55

56 _____

57 $^{59}_{27}Co$ + _____ $\rightarrow$ $^{60}_{27}Co$

58 _____

59 _____ g

60 _____ **min**

61 _____

_____

_____

62 Potential energy: _____

Average kinetic energy: _____

63 _____ **mol**

64 Change in pressure: _____

Change in temperature: _____

65

Part C

66 _____

67 _____

68 _____

69

70 _____

71 _____

_____

_____

72 _____

_____

_____

73 Rate of forward reaction: _____

Rate of reverse reaction: _____

74 _____

75 _____

76

```
      H   H   H   H   H   H   H   H   H   H   H   H   H   H   H   H   H   O
      |   |   |   |   |   |   |   |   |   |   |   |   |   |   |   |   |   ||
 H─C─C─C─C─C─C=C─C─C=C─C─C─C─C─C─C─C─C─C─O─H
      |   |   |   |   |       |           |   |   |   |   |   |   |
      H   H   H   H   H       H           H   H   H   H   H   H   H
```

77 _____H$_2$(g) + _____ O$_2$(g) → _____ H$_2$O(g) + energy

78 From _____ to _____

79 _____ **mol**

80 _____

81 _____ M

82 _____ mL

83

_____

_____

84 _____

85 _____